The Biology of Cell Reproduction

Renato Baserga

The Biology of Cell Reproduction

HARVARD UNIVERSITY PRESS

Cambridge, Massachusetts / London, England 1985

Printed in the United States of America
10 9 8 7 6 5 4 3 2 1

This book is printed on acid-free paper, and its binding materials have been chosen for strength and durability.

Library of Congress Cataloging in Publication Data

Baserga, Renato.
The biology of cell reproduction.

Bibliography: p.
Includes index.
1. Cell proliferation. 2. Cell cycle. I. Title.
[DNLM: 1. Cell Cycle. 2. Cell Division. 3. Cell Transformation. Neoplastic. 4. Molecular Biology. QZ 202 B299b]
QH605.B327 1985 574.87'62 84-12902
ISBN 0-674-07406-8 (alk. paper)

To my wife, Beverly

O]ἰ μὲν ἰππήων στρότον, οἰ δὲ πέσδων,
οἰ δε ναων φαῖσ' ἐπ[ι] γᾶν μέλαι[ν]αν
ἔ̆]μμεναι καλλιστον· ἔ̆γω δὲ κῆν' ὄ̆τ-
τω τις ἔ̆ραται.

Preface

Biologists have been interested in cell reproduction since the discovery of cell division in 1826 and Virchow's subsequent statement "Omnis cellula e cellula." The question "how do cells divide?" has been approached at different levels, first at a morphological level, then biochemically; but it has only been in the past 30 years, with the advent of modern cell biology and molecular biology, that the study of cell reproduction has made rapid progress. And this is what my book is about; it is an interdisciplinary approach to the question: how do cells divide? What I hope to have achieved is a book that provides molecular biologists with the basic information on the reproduction of animal cells, and that introduces cell biologists and clinicians to oncogenes and the molecular biology of the cell cycle.

It may seem presumptuous for a single author to cover all these different aspects of cell proliferation. As my justification, I can say that I began my research more than 30 years ago, studying experimental metastases in mice, and I am now cloning cell cycle genes, having gone through the intermediate stages of biochemistry and cell biology. I have therefore had first-hand experience with several aspects of cell reproduction, and my knowledge of the literature is derived from original articles, not just reviews (although in this book, to limit the number of references to a reasonable level, I have sometimes used review articles or books as the sources of the citations).

Students should find this book a good general introduction to the biology of cell reproduction, although most of the material refers to animal cells, with a few fleeting references to yeasts and none at all to bacteria or plants. The reason, again, was to keep the length of the book within reasonable limits.

I state in chapter 4 that we should console ourselves for the loss of brain cells that occurs with aging by remembering that, besides losing neurons containing information, we are also losing neurons packed with misinformation. The same comment applies to a book. Any book (and I have had a love affair with books from the time I learned to read) contains information and misinformation. All that an author can hope is that the amount of information is in large excess. It is up to the reader to apply the rule of the apostle Paul: "Omnia discite, ritenete bonum."

Finally, I would like to thank a number of colleagues who have freely given of their time to offer suggestions, criticisms, and encouragement during the four years it took me to prepare this book: Norbel Galanti (University of Chile, Santiago, Chile), Beverly Lange (Children's Hospital of Philadelphia, Pennsylvania), Charles Stiles (Dana-Farber Cancer Institute, Boston, Massachusetts), Leszek Kaczmarek (Memorial Hospital Child Health Center, Warsaw, Poland), and, at Temple University, Ricky R. Hirschhorn, W. Edward Mercer, Susan Rittling, and John Wurzel.

Contents

Part I

Biology

Chapter 1

The Cell Cycle

For over a century we have known that the basic mechanism of reproduction in metazoan cells is mitotic division. Although growth in size of individual cells plays a role in the growth of organs and tissues, an increase in the number of cells is by far the most important component of both normal and abnormal growth. Since this increase in cell number is due to mitotic division of cells, an understanding of normal growth and its derangements must be based on a study of the mechanisms that regulate cell division.

Until recently, very little was known about the biochemical or molecular events that precede and regulate mitosis. Although scientists had long been aware that dividing cells have lengthy intervals of apparent rest between one mitosis and the next, only in the past few years have they realized that this interval, called the interphase, is not really a period of rest but is, on the contrary, a period of intense activity during which the cell prepares for mitosis. This book will consider in detail these events; but in order to better understand the biochemical and molecular basis of cell proliferation, it is first convenient to examine the behavior of cells as individuals and as populations. We shall therefore discuss in Parts I and II the biology of cell division and the way single cells increase in size and populations of cells increase in number and, thus, cause the growth of organs and tissues. Then in Part III we will consider the growth factors, the genes and gene products that are involved in the proliferation of eukaryotic cells.

We shall begin our study of cell reproduction in metazoan cells with a description of the cell cycle. A knowledge of the cell cycle will not only give us the biological basis through which populations of cells grow in number but, even more important, will provide us with a framework to sort out the biochemical and molecular events relevant to cell division.

PHASES OF THE CELL CYCLE

The cell cycle is defined as the interval between completion of mitosis in the parent cell and completion of the next mitosis in one or both daughter cells. It, therefore, comprises both the interphase and the mitosis of the old textbooks of biology. Recent methodology has allowed us to divide the intermitotic interval into four phases (Fig. 1.1): presynthetic interphase or postmitotic gap (G_1), DNA synthesis phase (S), postsynthetic interphase or premitotic gap (G_2), and mitosis (M). The subdivision of the intermitotic interval into four phases and the recognition of the cell cycle were made possible by the development of new methods, both in vivo and in vitro. Among these the two most common methods for investigating the cell cycle are autoradiography and cytophotometry. A description of the methodologies used will make it easier to understand the meaning of the four phases of the cell cycle and will also be useful in studying the growth behavior of populations of cells.

The Cell Cycle by Autoradiography

Direct observation by light microscopy distinguishes cells in mitosis from interphase cells; autoradiography distinguishes cells syn-

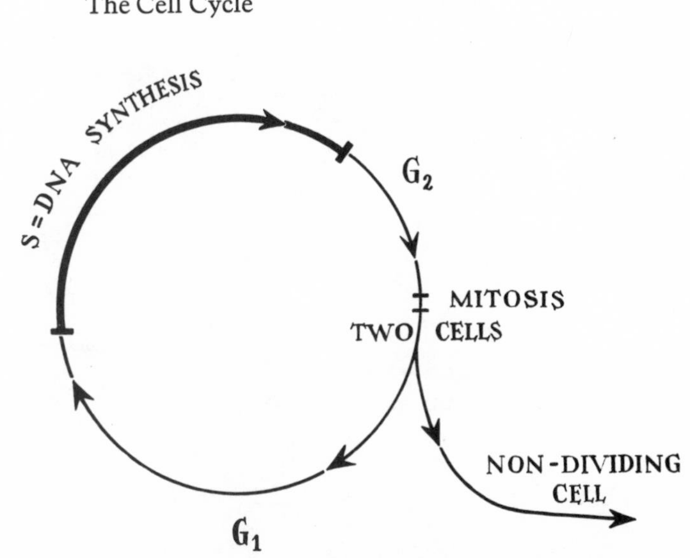

Fig. 1.1 Schematic diagram of the cell cycle. After mitosis, the progeny cells enter the G_1 phase. This gap period is preparatory to a second phase, the S phase (for synthesis), a discrete period of the cell cycle during which DNA synthesis and indeed replication of the whole genome occur. After completion of DNA synthesis, the cells enter a second gap period (G_2), preparatory to mitosis. After a certain number of divisions, some cells leave the cycle and become nondividing, terminally differentiated cells.

thesizing DNA from cells which are not (Fig. 1.2). Thus, by simply looking through a microscope at autoradiographs one can divide a population of cells into three categories: mitotic cells, cells synthesizing DNA, and cells that are not synthesizing DNA and are not in mitosis. Although the original experiment in which Howard and Pelc (1951) first described the cell cycle was done with [^{32}P], a much better label for cells synthesizing DNA is [^{3}H]-thymidine. It has three advantages: thymidine is incorporated exclusively into DNA, nonincorporated thymidine is washed out by ordinary fixatives used to fix tissues and smears, and the tritium label allows high resolution autoradiography (Fig. 1.2).

The cell cycle can be adequately portrayed by describing a typical experiment using autoradiography. Briefly, the method consists in labeling a block of cells (those synthesizing DNA) and watching that block of labeled cells as it moves past a fixed point in the cell cycle, mitosis (Fig. 1.3). As an illustration, let us take replicate cultures of cells plated on coverslips at a concentration of 10^4 cells/cm^2 in an appropriate growth medium. On the second day after plating, when the cells are growing exponentially, all cultures are pulse-labeled

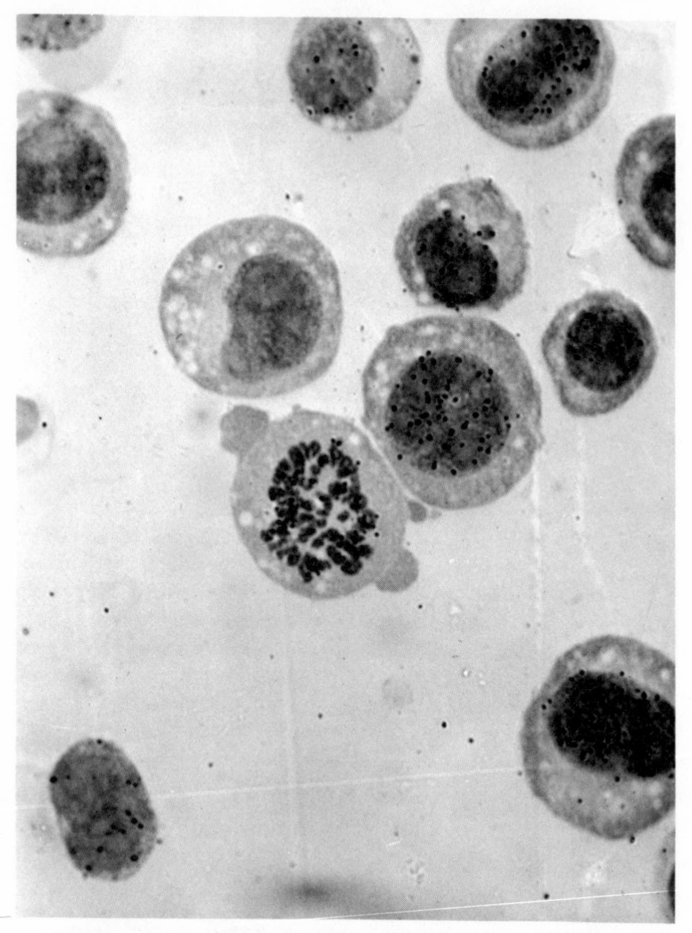

Fig. 1.2 Autoradiography of cells exposed to [^{3}H]-thymidine. The cells were labeled with [^{3}H]-thymidine for 30 minutes and fixed a few hours later. Cells that were synthesizing DNA at the time of exposure to the radioactive precursor are labeled, and the label is identified by the presence of silver grains in the emulsion overlying the cells. Since thymidine is exclusively incorporated into DNA, the [^{3}H] label is confined to the nuclei. In fact, the mitotic cell in the center shows that the label follows the distribution of the chromosomes.

with [^{3}H]-thymidine, i.e., they are exposed for 20 minutes to [^{3}H]-thymidine, 0.1 μCi/ml. After labeling, all cultures are washed carefully with balanced salt solution to remove excess thymidine and then reincubated in nonradioactive growth medium. At various intervals after pulse labeling, two or three cultures are terminated, and

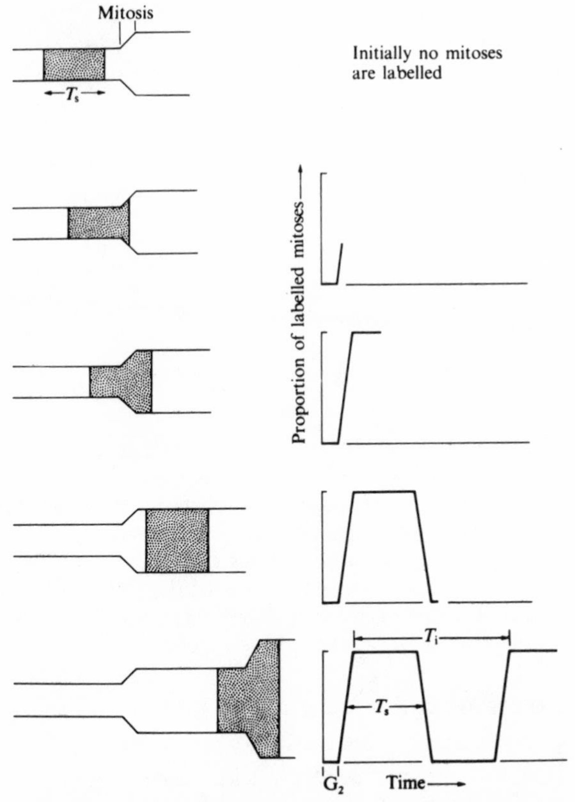

Fig. 1.3 Cell cycle analysis by the technique of labeled mitoses. After a short pulse of [^{3}H]-thymidine, the labeled cohort of cells (the stippled area) moves through mitosis. As it does so, the proportion of labeled mitoses describes a peak, whose width is equal to the duration of S phase (T_S). After a suitable interval, the proportion of labeled mitoses peaks again. The interval between the two peaks is the cell cycle time (T_i in this figure, usually written T_c). The length of G_2 is also determined experimentally, while G_1 is usually calculated as $G_1 = T_c - (T_S + T_{G_2} + T_M)$. (Reprinted, with permission, from Steel 1977.)

the cells on the coverslips are fixed in Carnoy's, or another fixative, and autoradiographed. Fig. 1.4 summarizes the results of such an experiment.

When cultures are terminated immediately after labeling, no labeled mitoses are found, while 30 – 35 percent of the interphase cells are labeled. This already tells us that DNA is synthesized in interphase but not during mitosis. Two hours after removal of [^{3}H]-thy-

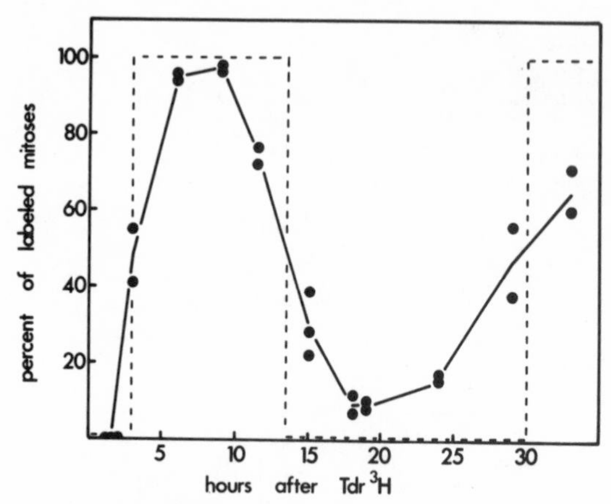

Fig. 1.4 Percentage of labeled mitoses in HeLa cells as a function of time after a 30-minute pulse of [^{3}H]-thymidine. The broken line represents the theoretical curve that would obtain if there were no individual variation among cells in the various phases of the cell cycle. The solid curve follows the actual data (closed circles). The calculation of the duration of the cell cycle and its phases from this curve is explained in the text. (Reprinted, with permission, from Baserga and Wiebel 1969.)

midine, mitoses are still unlabeled; and since mitoses last only 30–45 minutes (see below), there must be a period before mitosis during which cells do not synthesize DNA. The length of this period, called the G_2 phase, is the interval between the time of exposure to [^{3}H]-thymidine and the time when 50 percent of the mitoses are labeled (3.5 hours in Fig. 1.4).

As the time interval between pulse-labeling and termination of cultures increases, the percentage of labeled mitoses increases rapidly to 100 percent. These mitotic cells are all cells that were synthesizing DNA when the cultures were exposed to [^{3}H]-thymidine. Then, as other cells that were *not* synthesizing DNA at that time enter mitosis, the percentage of labeled mitoses drops. Because of variations in cell cycle times among individual cells, the curve of percentage of labeled mitoses in Fig. 1.4 has the shape of the curve with closed circles. If there were no variations, the percentage of labeled mitoses would increase almost instantaneously from 0 to 100, remain at 100 percent for the entire length of the DNA synthesis period, and then suddenly decrease from 100 to 0 (dashed line in Fig. 1.4). Instead, the curve slopes upward at the beginning and, after a few hours at or near 100 percent, decreases gently. After reaching a low point (but very rarely zero because of the variation among individual cells), the percentage of labeled mitoses increases again, as progeny of the cells that were in mitosis 3–4 hours after pulse-labeling in turn enter mitosis.

If you imagine yourself standing at the mitotic post and watching labeled cells go by, you can translate the data of Fig. 1.4 into the diagram of Fig. 1.1. The phases of the cell cycle are deduced and their lengths calculated as follows:

> G_2 phase (postsynthetic or premitotic gap) = interval between removal of label and time when 50 percent of the mitoses are labeled. In Fig. 1.4: 3.5 hours.
>
> S phase (DNA synthesis phase) = interval between the 50 percent points on the first ascending and descending limbs of the curve of percentage labeled mitoses. In Fig. 1.4: 10 hours.
>
> G_1 phase (presynthetic or postmitotic gap) = interval between 50 percent points on the first descending and the second ascending limbs. In Fig. 1.4: ~13 hours.
>
> Mitosis is visible by light microscopy. The duration of mitosis can be calculated to be ~0.5 hours, either by autoradiography (Fry, Lesher, and Kohn 1962) or, better, by time lapse cinematography (Sisken and Wilkes 1967).
>
> Cell cycle time (intermitotic time, T_c) = interval between two mitoses, i.e., interval between two corresponding points on the first and second ascending limbs of the curve of percentage of labeled mitoses. In Fig. 1.4: 27 hours.

There are several other ways by which autoradiography can be used to calculate the length of the cell cycle and its phases, but the labeled mitosis curve is the most popular and probably the most informative. Several good books give the technical details of autoradiography and discuss the various methods of cell cycle analysis, but, of course, I am partial to the one by Baserga and Malamud (1969). Despite its venerable age, it is still useful because autoradiography has not changed much in the past 15 years and, at any rate, the other good books on autoradiographic methods are approximately of the same vintage. The same analysis of the cell cycle can be carried out in whole animals. The pulse-labeling is automatic in vivo because the thymidine that is not incorporated into DNA is rapidly broken down, largely by the liver, into nonutilizable products. A single injection of [^{3}H]-thymidine into an animal is therefore equivalent to a pulse-labeling in cell cultures, and the period of labeling is estimated to last 30–45 minutes.

Tissues of choice for studying the cell cycle in animals are those that have actively dividing cell populations, such as the lining epithelium of the small and large bowels, the bone marrow, the skin and mucosae, and transplantable tumors. Of course, in these experiments animals have to be killed at different intervals after injection of [^{3}H]-thymidine. One of the classic papers on the cell cycle is the one by Quastler and Sherman (1959) on cell population kinetics in the intestinal epithelium of mice. It introduced the concept of a progenitor compartment where cell proliferation occurs (in the case of intestinal epithelium, the crypt cells) and a functional compartment, where cell proliferation has ceased. It even introduced the concept of a stem cell, two ideas to which we shall return in a later section. For the acumen in the construction of a theoretical cell cycle model, this paper is unsurpassed. Interestingly enough, their estimates of the length of the phases of the cell cycle were inaccurate, being overestimated by about 40 percent.

The Cell Cycle by Cytophotometry

One of the early landmarks of eukaryotic molecular biology was the report, in 1948, by Boivin, Vendrely, and Vendrely that the amount of DNA per cell was constant. These authors measured the average DNA content of nuclei in different tissues of cattle and found 6.6×10^{-12} g DNA per cell in all tissues except in sperm, where the amount, 3.3×10^{-12} g, was exactly half. Back in 1948, this was a remarkable finding, because it indicated that all cells contain the same amount of genetic information, and that differences between cells must depend on gene expression rather than on the presence or absence of genes. As it happens often with seminal findings, the details were wrong but the principle remains. It is true that most somatic cells contain the diploid or 2n amount of DNA. However, as we shall see later, a substantial fraction of somatic cells contain a 4n (tetraploid) amount of DNA, and even 8n or more. In normal nondividing somatic cells, though, the amount of DNA is always a precise multiple of 2n, and these are called euploid cells. Some cancer cells and most cells in culture have an aneuploid amount of DNA, i.e., a DNA content that is not an exact multiple of 2n.

Cells in DNA synthesis will also have an intermediate amount of DNA—if diploid, $>2n < 4n$, or, since not all cells are diploid, an amount of DNA that is intermediate between the G_1 amount and G_2 amount. If one can measure the amount of DNA per cell, one can obtain a distribution of cells throughout the cell cycle. In the early studies on the cell cycle, the amount of DNA per cell was measured either by cytophotometry on fixed cells stained with a DNA specific dye, or by ultraviolet microspectrophotometry. A review of these early studies can be found in Vendrely (1971). Today, flow cytophotometry is the most popular methodology, although microspectrophotometry on fixed cells, aided by computers, has staged a remarkable comeback. Flow cytophotometry is faster and can be done on a large number of live cells; the instrumentation also allows the sorting of different cell populations. Let us have a look at two examples of cell cycle analysis by flow cytophotometry, specifically: (a) measurement of the amount of DNA; (b) simultaneous measurement of DNA and RNA amounts. For the details on the instrumentation and the history of flow cytophotometers and cell sorters, the reader is referred to the book by Melamed, Mullaney, and Mendelsohn (1979).

DNA Content by Flow Cytophotometry We will use as an illustration the paper by Tobey and Crissman (1975). CHO cells were fixed and stained with mithramycin or other dyes, which bind specifically to DNA. Following laser excitation, the cells fluoresce with an intensity proportional to DNA content. The results obtained with asynchronous exponentially growing CHO cells are shown in Fig. 1.5, which compares the DNA content distribution of cells stained either by the acriflavine-Feulgen procedure or with mithramycin. The first peak represents G_1 cells, the last peak G_2 cells, and the curve in between S phase cells. The percentage of cells in G_1, S, and $G_2 + M$ can be obtained by computer-fit analysis of the DNA distribution curves. From these percentages, if one knows T_c (which, however, must be obtained by other methods), one can determine the length of the various phases of the cell cycle. A comparison with the autoradiographic method is useful. Flow cytometry is much faster, does not require the use of radioactive compounds (very important in in vivo studies), and is ideal for studying perturbations of the cell cycle. Autoradiography gives additional information (morphology of the cells, growth fraction) but is time consuming.

Many other dyes have been used besides mithramycin and Feulgen

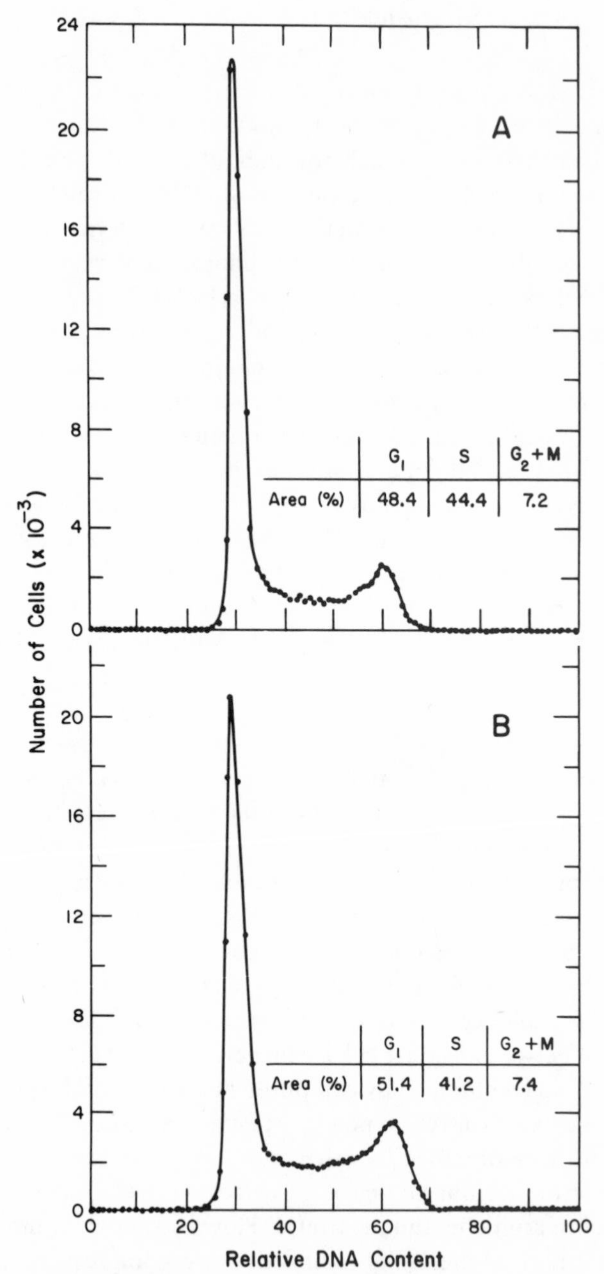

Fig. 1.5 DNA distribution of exponentially growing Chinese hamster cells in culture. The cells were stained either by the acriflavine-Feulgen procedure (A) or with mithramycin (B) and the amount of DNA measured by flow cytophotometry. The percentages of cells in the different phases of the cell cycle were obtained by computer-fit analyses of the DNA distribution curves. (Reprinted, with permission, from Tobey and Crissman 1975.)

Table 1.1. Cell cycle analysis derived from DNA distributions in two cell types stained with various fluorochromes.

Stain	Cell cycle distribution (%)		
	G_1	S	$G_2 + M$
		WI-38 CELLS	
Acriflavine	48.9	37.5	13.6
Flavophosphine	53.5	33.0	13.5
Ethidium bromide	48.7	38.3	13.0
Propidium iodide	49.4	36.6	14.0
		HeLa CELLS	
Acriflavine	54.7	38.1	7.2
Flavophosphine	53.6	38.6	7.8
Ethidium bromide	52.8	36.3	10.9
Propidium iodide	52.3	37.9	9.8

Adapted from Crissman et al. (1979). The percentages of cell cycle phases were obtained by computer analysis of DNA distributions.

staining. These include ethidium bromide, propidium iodide, acriflavine, flavophosphine, chromomycin A3, olivomycin, Hoechst 33258, Hoechst 33342, and others. These are discussed by Crissman et al. (1979); from their paper I have adapted Table 1.1 summarizing the DNA distributions of two types of cells using different fluorochromes. Except for flavophosphine in WI-38 cells, different fluorochromes give comparable values. The S phase may be slightly underrepresented in comparison to values obtained by autoradiography. This is probably due to the fact that cells in very early or late S phase have DNA amounts that are easily confused with G_1 or G_2 amounts, respectively.

Simultaneous Measurement of DNA and RNA Amounts In this technique, cells in suspension are stained with the fluorescent metachromatic dye acridine orange under conditions such that the dye intercalates into native DNA and fluoresces orthochromatically green with maximum emission at 530 nm (F_{530}) while RNA stains metachromatically red with maximum emission at 640 nm ($F_{>600}$).

Fluorescence of individual cells is measured in a flow cytophotometer. Red and green fluorescence emissions from each cell are separated optically and the integrated values of the pulses quantitated by separate photomultipliers (Darzynkiewicz et al. 1976). Under appropriate conditions, green fluorescence is directly proportional to DNA content, and red fluorescence is a reasonable measurement of RNA content (Darzynkiewicz et al. 1977).

The results can be presented in several ways. Fig. 1.6 shows a computer-drawn scattergram of the distribution of individual cells according to their green and red fluorescence intensities after they are stained with acridine orange. Each dot is an individual cell with given amounts of RNA and DNA. The advantage of this technique is that it can correlate the two amounts (of course, one can stain for

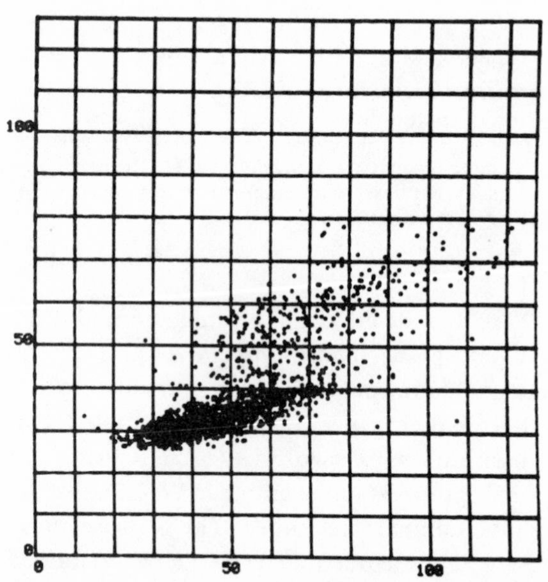

Fig. 1.6 Computer-drawn scattergram of tsAF8 cells stained with acridine orange and analyzed by flow cytophotometry. Each dot represents a single cell. The abscissa gives the amount of RNA per cell (expressed in arbitrary units of red fluorescence) and the ordinate the amount of DNA per cell (expressed in arbitrary units of green fluorescence). In this scattergram, cells with <40 units of green fluorescence have a G_1 content of DNA; cells with >70 units are G_2 cells; and those with intermediate values are in S phase. Note the increase in RNA amounts as the cell progresses through the cell cycle.

proteins instead of RNA). Needless to say, it gives different information than bulk chemical analysis for RNA and DNA, although a few years ago I had a hard time convincing a journal editor of this. (I mention it simply as a comfort to all of us whose papers are returned by journal editors and as a reminder that reviewers are not necessarily wiser than authors.) Eventually, I had to insert in the paper a statement that should be obvious to any high-school student, namely, that in bulk determinations a 30 percent increase in RNA amount could be due to a 30 percent increase in 100 percent of the cells or a 100 percent increase in 30 percent of the cells. Notice from Fig. 1.6 how the flow cytophotometer instead assigns to each cell their specific RNA and DNA contents. These data can also be computed and displayed on frequency distribution histograms (Darzynkiewicz et al. 1980), examples of which are given in Fig. 1.7. The histogram for green fluorescence is the same as an ordinary histogram obtained with mithramycin or propidium iodide. Notice the wide distribution of red fluorescence intensities. Finally the data can also be presented in computer-drawn two-parameter frequency histograms (Darzynkiewicz et al. 1980).

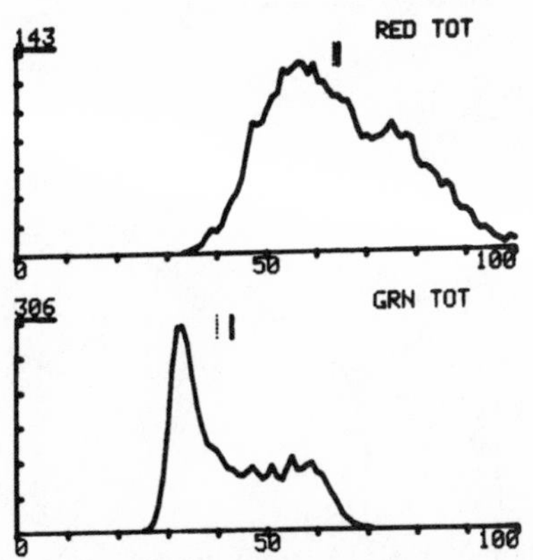

Fig. 1.7 Frequency distribution histograms of cells stained with acridine orange and analyzed by flow cytophotometry. Top: distribution of red fluorescences (RNA). Bottom: distribution of green fluorescences (DNA). The abscissa gives arbitrary units of either red or green fluorescence, the ordinate the number of cells. The DNA distribution is similar to that in Fig. 1.5. Notice the broader distribution of RNA amounts.

DNA Content by Microspectrophotometry Computer-operated microspectrophotometers can also be used in place of flow cytophotometry. Cells, on slides or coverslips, can be stained with acridine orange or other dyes as described above and the amount of fluorescence (or color intensity) determined quantitatively. An illustration is given in Fig. 1.8. The advantages of computer-operated microspectrophotometers include: (1) the cells are stained on a coverslip, avoiding the trypsinization that often damages cells in monolayer cultures and in solid tissues; (2) the cells are viewed with a light microscope and can be selected for measurement, a notable aid in cytology smears and in tissues sections, which are a mixture of different cell types; (3) nuclear and nucleolar antigens can be quanti-

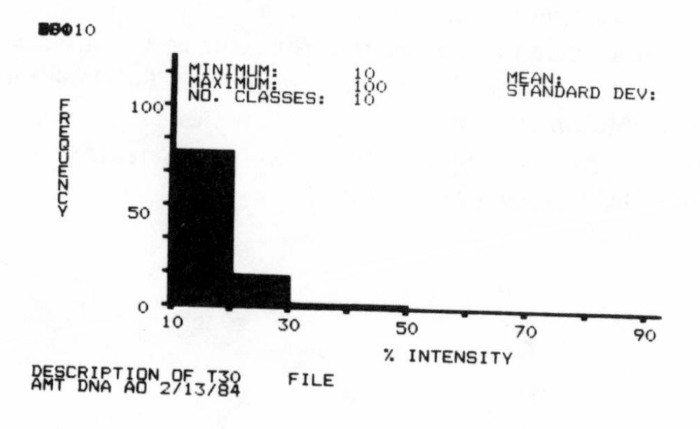

CLASS LISTING

CLASS	INTERVAL FROM	TO	ABS. COUNT	REL. %
1	10	20	78	80.4
2	20	30	16	16.4
3	30	40	2	2
4	40	50	1	1
5	50	60	0	0
6	60	70	0	0
7	70	80	0	0
8	80	90	0	0
9	90	100	0	0
10	100	110	0	0

Fig. 1.8 Histograms of DNA amounts in quiescent 3T3 cells. The amount of DNA per cell was determined by measuring the green fluorescence of acridine-orange–stained cells using a computer-operated microspectrofluorimeter. Fluorescence intensity is expressed in arbitrary units (abscissa); frequency is on the ordinate. The figure is the actual print-out given by the computer-operated microspectrofluorimeter. Cells with <30 arbitrary units (97%) have a G_1 content of DNA. Three cells have an S phase DNA content (30–50 arbitrary units).

tated quite accurately; and (4) a small number of cells is required, ideal for experiments using microinjection or transfection.

The three methods of determining DNA metabolism (autoradiography, flow cytophotometry, and computer-operated microspectrofluorimetry) are not mutually exclusive. On the contrary, they complement each other and can sometimes be used even simultaneously. In my laboratory, we favor computer-operated microspectrofluorimetry, because the instrumentation is more flexible and gives more information than the flow cytophotometer. It is true that the microspectrofluorimeter has no sorting capacity, but I have heard from many colleagues that the sorting capacity of many flow cytophotometers (unless specially modified) is unreliable. Regardless of the instrumentation used, the data generated by all these methods can give us an accurate kinetic picture of the cell cycle and its phases, as well as valuable information on the size and morphology of cells.

VARIABILITY OF THE CELL CYCLE

The average length of the cell cycle and its phases varies from cell type to cell type; even in the same cell type, it depends upon different physiological conditions (age, growth factors, hormones, etc.). In addition, in a homogenous population of cells, individual cells have variable cell cycle times. The number of cell types that have been characterized in terms of cell cycle parameters (length of cycle, G_1, S, G_2, and mitosis) is such as to preclude a complete, or even a partial, listing. Just to give a rough idea, I have summarized in Table 1.2 some examples of representative cell cycle parameters. For reasons of space, it is not possible to list every reference from which the data that contributed to Table 1.2 were taken, and I only give the reviews and books in which these data have been summarized. The reader who wishes to know more about a specific cell type will have to consult these reviews and books to find the original source.

A few generalizations on cell cycle parameters can be made, although exceptions are frequent. (1) Mitoses usually last less than one hour, although their duration can sharply increase in tumors and transformed cells; (2) the length of G_2 varies from 2–4 hours, and again it is often increased in tumor cells; (3) the length of S phase, in the majority of cases, is reasonably constant, ~7–8 hours, although exceptions are often observed; (4) the most variable phase of the cell cycle is the G_1 period. This is also easily observed in cells of the same

Table 1.2. Cell cycle times.

Cell Type	T_c	T_{G_1}	T_S	$T_{G_2}+M$
CELLS IN CULTURE				
HeLa S3	21	8.0	9.5	3.5
Human diploid fibroblasts	18	6.5	7.5	4.0
KB cells	31.5	6.5	7.5	17.5
Human amnion cells	19.5	9.5	7.0	3.0
L mouse fibroblasts	18	8.0	6.0	4.0
3T3	19	8.0	7.0	4.0
Chinese hamster ovary cells	14	5.5	4.5	4.0
Chinese hamster lung cells	10	1.5	6.0	2.5
MAN IN VIVO				
Colon epithelial cells	25	9	14	2
Rectum epithelial cells	48	33	10	5
Stomach epithelial cells	24	9	12	3
Bone marrow cells	18	2	12	4
Basal cell carcinoma	67	36	19	12
Epidermoid carcinoma	24	9	11	4
Acute myeloblastic leukemia	49	24	21	4
Melanoma	46	20	19	7
Ascites cells from carcinoma	113	50	48	15
OTHER ANIMALS IN VIVO				
Mouse				
Duodenal epithelium	10.3	1.3	7.5	1.5
Ileum crypt cells	10.1	1.8	6.9	1.4
Colonic epithelium	19	9	8	2
Growing hair follicles	12	3	7	2
Mammary gland, alveoli	71	45	22	4
Same after hormonal stimuli	13	1.3	9.2	2.5
B16 melanoma	16.5	5.3	8.3	2.9
Lewis lung carcinoma	17.6	5.0	9.6	3.0
Ehrlich ascites tumor	16.4	3.1	10.0	3.3
Rat				
Duodenal crypt cells	10.4	2.2	7.0	1.2
Liver cells (8 weeks old)	47.5	28.0	16.0	3.5
Internal enamel epithelium	27.3	16.0	8.0	3.3
Hepatoma cells	24	12.7	7.9	3.4
Hamster melanoma	16	9.6	4.8	1.6

Compiled from sources cited in Baserga and Wiebel (1969), Baserga (1976), and Steel (1977). Mitosis is arbitrarily calculated to last 1 hr. T_c = length of the cell cycle; T_{G_1} = length of G_1 phase, etc. All times in hours.

type, using synchronized cultures: for instance, in HeLa cells, the lengths of S, G_2, and M are constant, but the length of G_1 varies in individual cells (Terasima and Tolmach 1963). In fact, it has been known for a long time that some cell lines do not even have a detectable G_1 period. Embryo cells in the first divisions postfertilization lack a G_1 period (Graham and Morgan 1966), but then their whole cell cycle is only a few minutes long. The slime mold *Physarum polycephalum*, Ehrlich ascites cells, and the micronuclei of *Tetrahymena* are all G_1^- (for a review see Malamud 1971). But only recently have the elegant experiments of Prescott and co-workers clearly pointed out the significance of a missing G_1 period (Liskay and Prescott 1978, Liskay 1978, Stancel, Prescott, and Liskay 1981).

V79 – 8 cells are a Chinese hamster cell line that lacks G_1, i.e., they are G_1^-. From it, mutants with a G_1 period (G_1^+) can be derived. When G_1^- cells are fused with G_1^+ cells, all the resulting hybrids are G_1^-, indicating that the G_1^- state is dominant. Complementation tests among different G_1^+ mutants showed that some of these mutants can complement, suggesting that they are G_1^+ for different reasons. Prescott and co-workers concluded that cells have a G_1 only when at completion of mitosis something is missing that is necessary for cell growth. Indeed, most of the G_1 period in G_1^+ hamster cells can be eliminated by slowing down growth with cycloheximide or by lengthening the S period with hydroxyurea. From this point of view, G_1 is not a phase of the cell cycle with specific biochemical events but only a period of time between mitosis and S phase. The conclusion of Stancel, Prescott, and Liskay (1981) is worth quoting: "We propose that the G_1 period in cultured cells is not part of the chromosome cycle but belongs to the growth cycle. If doubling in cell size is completed as rapidly as the chromosome cycle (equal to $S + G_2 + M$) the cell cycle will lack a G_1 period. When growth is slower, the initiation of DNA synthesis is delayed (a G_1 period is introduced) until growth is completed." We shall return to this duality of growth in size and cell DNA replication in chapter 9. For the moment, let us concentrate on the hypothesis of Prescott and co-workers that G_1 is not part of the chromosome cycle.

Although the authors themselves recognize that this may not be true in vivo, the concept of G_1 as a period of time is a fruitful one. Even if entry into S does not depend on cellular size (see chapter 11), it is evident that a cell could complete its requirements for re-entry into S by the end of mitosis. This much said, G_1 remains a convenient notation to indicate cells and populations of cells that have completed mitosis but are a few hours away from the beginning of S. In other words, G_1 is dispensable, but some cells *are* in G_1, and the length of the cell cycle is largely regulated by lengthening G_1. Indeed, Darzynkiewicz et al. (1980), on the basis of results from flow cytometry, have proposed that the G_1 period can be subdivided into two subcompartments, which they term G_{1A} and G_{1B}. According to these authors, only G_{1B} cells are competent to enter DNA synthesis.

Even in a well-synchronized population of cells in culture, the length of G_1 is highly variable from cell to cell despite the fact that all cells have started together from mitosis and are all in the same controlled environment. What causes this variability? There is as yet no answer to this question, but I would like to propose an explanation, while acknowledging that there is not a single shred of evidence for it. I suggest that mitoses are often asymmetrical in terms of cytoplasmic components, and that there is unequal distribution of mRNAs and proteins to the daughter cells. As we shall see later, unique copy gene transcription is necessary for the $M \rightarrow S$ transition, and the length of G_1 may be determined by the number and amounts of certain specific mRNAs and proteins that the cell has to accumulate for entry into S. Interestingly enough, in the developing *Xenopus* embryo, a G_1 period appears at the same time as de novo RNA synthesis (Newport and Kirschner 1982).

One more observation on Table 1.2. Notice that the cell cycle time of some normal cells is actually shorter than the cell cycle time of the fastest growing tumors. This seemingly paradoxical situation will be dealt with in chapter 4. However, even at this point the reader may wish to consult a review by Bresciani (1968) in which the cell cycles of normal and tumor tissues are compared and discussed.

Cell cycle times can be manipulated by hormones, concentrations of growth factors and nutrients, and drugs. Each of these manipulations will be dealt with in the respective chapters. I will mention here, though, that in tissue cultures, the cell cycle can be affected by temperature. The length of the cell cycle changes very little between 36° and 39°, but it increases below 36° or above 39°. Whereas only G_1 varies with other factors that can affect the length of the cell cycle, with high or low temperatures all phases of the cell cycle are proportionally changed (Sisken, Morasca, and Kibby 1965).

Finally, there is another source of variability in kinetic studies of the cell cycle in animals, and that is a circadian rhythm. Pilgrim, Erb, and Maurer (1963) were the first to show a circadian rhythm in the mitotic index and the thymidine labeling index in various mouse tissues, such as the epithelia of esophagus, fore-stomach, tongue, and epidermis. The labeling index (i.e., the fraction of cells labeled by a pulse of [^{3}H]-thymidine) reached a daily peak at 17.30 and a minimum at 11.30, while the mitotic index was highest between 17.30 and 23.30. For instance, in the tongue epithelium, the labeling index was 20.7 percent at 17.30 and 4.3 percent at 11.30. A circadian rhythm can also be detected in rapidly proliferating tissues, such as the epithelium and the pericryptal fibroblasts of the small intestine and colon of mouse (Neal and Potten 1981).

Chapter 2

Nondividing Cells

Thus far we have discussed only dividing cells, since the cell cycle (the subject of chapter 1) is defined as the interval between completion of mitosis in the parent cell and completion of the next mitosis in one or both daughter cells. There are indeed populations of cells in which all cells are dividing. In vitro, this is true of most cell lines during their exponential phase of growth. When exponentially growing cells are labeled with [^{3}H]-thymidine for 24 hours, it is not uncommon to find that >99 percent of cells become labeled. In vivo, in the earliest stages of embryonal development immediately after the fertilization of the egg, all cells divide. However, most populations of cells consist of a mixture of both dividing and nondividing cells. The nondividing cells can be of two types: G_0 cells, that are still capable of re-entering the cell cycle; and terminally differentiated cells, that are destined to die without dividing.

G_0 CELLS

In Vitro

I define the G_0 state as a state of dormancy in respect to growth. In this state, the cells can perform their physiological functions (whatever they may be) and from this state they can be triggered into the proliferative phase by an appropriate stimulus.

We shall take as an illustration the original experiment of Todaro, Lazar, and Green (1965) with 3T3 cells. These cells during exponential growth have a doubling time of 18 hours; they slow down when they reach a density of about 2×10^4 cells/cm^2 and stop when the monolayer is complete, at a saturation density of about 5×10^4 cells/cm^2. Virtually no mitoses can be detected in these density-inhibited cultures and, if labeled with [^{3}H]-thymidine for 24 hours, at most 2 – 3 percent of the cells become labeled. In this state of quiescence, the cells can remain viable for several days. However, if the medium is now replaced with fresh medium containing 10 percent serum, the density-inhibited cells re-enter the cell cycle. First they enter the S phase and then they undergo mitosis. This means that the cells are arrested at some point *after* mitosis and *before* S phase. This point, though, is not a point in the G_1 phase of cycling cells. It is different and we call it G_0, although other terms like Q cells, quiescent cells, resting cells, etc., are used in the literature. We will have many opportunities to notice differences between G_0 and G_1 cells. For the moment, let me simply point out that the time interval between serum stimulation of G_0 cells and the beginning of S phase is longer than the time interval between mitosis and S. For instance in WI-38 cells, G_1 is 6 hours, but in G_0 cells the interval between serum stimulation and entry into S — which I like to call the pre-replicative phase — averages 18 hours. Indeed, even in cell fusion experiments, G_0 nuclei are much slower than G_1 nuclei in responding to inducers of DNA synthesis (Mercer and Schlegel 1980; Rao and Smith 1981). So, G_0 emerges as a physiological state of the cell that the cell enters after mitosis and from which it can be rescued by an appropriate stimulus.

Under optimal conditions, almost all G_0 cells can be induced to re-enter S phase after a lag period. The addition of serum is not the only means for releasing density-inhibited cells from G_0. When a confluent culture of 3T3 cells, in which most cells do not synthesize DNA, is wounded, DNA synthesis is initiated in the cells that migrate into the area denuded by wounding (Todaro, Lazar, and Green 1965; Dulbecco 1970). Wounding simply consists of removing a strip of cells, for instance with a spatula, from the surface of the culture dish. Because the requirements for release from G_0 differ slightly in serum stimulation and in wounding, the term "topoinhibition" has been used to describe the inhibition of cell proliferation caused by extensive cell-to-cell contact (Dulbecco 1970). The difference, subtle but important, is in the type of release rather than in the type of inhibition. In density-inhibited serum-stimulated cultures, it is the addition of new growth factors that causes them to enter S; the cells remain in contact with one another. In topoinhibition, the cells are released by the breaking of cell-to-cell contact, not by the addition of growth factors. Another distinction has been proposed, with some merit, by Dethlefsen(1980), who would like to restrict the term G_0 "to viable cells that are out of cycle under normal physiological conditions (i.e., not nutrient deprivation per se) and can be recruited into active proliferation by a proper stimulus." Dethlefsen considers G_0 cells to be a subset of a larger population of noncycling cells, which he calls Q cells, that includes all cells out of cycle irrespective of the underlying mechanism. I would tend to agree with his distinction, provided growth factors are *not* included among nutrients.

Not all kinds of cells in culture can enter the G_0 state. The most striking example are cells transformed by either SV40 or polyoma, but many other cell lines (for instance, HeLa cells and most CHO cell lines) cannot be arrested in G_0.

The G_0 state itself varies with the time the cells have been quiescent. For instance, when WI-38, a strain of human diploid fibroblasts, are plated at a density of 2×10^4 cells/cm^2, they become confluent and quiescent within 5 days. The cultures display few mitotic figures and only 2.7 percent of the cells are labeled by a 24-hr pulse of[^{3}H]-thymidine. The longer the cells are left confluent without medium change, the lower is the number of mitoses and of DNA synthesizing cells. More important, when quiescent cultures are serum-stimulated, the longer the cells have remained quiescent, the longer is the interval between addition of fresh serum and the time of re-entry into S. Thus, confluent cultures stimulated on the fifth day after plating enter S phase after a lag of 10 hours (average). This average lag period increases to 16 hours in cultures stimulated on the ninth day after plating and to 24 hours in cells stimulated 18 days after plating (Augenlicht and Baserga 1974). However, as pointed out by Miska and Bosmann (1980), there is an upper limit to the increase in the lengh of the pre-replicative phase with increasing time of quiescence.

To say that G_0 cells are in a toxic state is incorrect. Cells in G_0 are not deprived of nutrients and their metabolic functions are not inhibited; they are simply deprived of growth factors. And to say that G_0 is an artifact of tissue cultures neglects the obvious fact that tissue culture itself is artifactual. Indeed, we will see below that the G_0 state is even more convincing in vivo. It suffices to say here that G_0 cells are viable and that virtually all of them can re-enter the cycle under appropriate conditions. Thus, we can conclude that, in tissue cultures, under conditions restrictive for growth (essentially a deficiency of growth factors), many cell lines arrest in a physiological state of quiescence, which we call G_0 and which is fully reversible. In chapter 10, in discussing the growth factors, we will study the countless ways in which G_0 cells in culture can be stimulated to re-enter the cell cycle.

Cells can be kept quiescent and viable for long periods of time. For instance, we have kept WI-38 cells quiescent either in 1 percent serum or in conditioned medium (i.e., medium obtained from cultures of WI-38 grown in 10 percent fetal calf serum for 7 days) for almost 100 days. When plated at a density of 2×10^4 cells/cm^2, WI-38 cells reached a maximum density of 6×10^4 cells/cm^2 by day 14, and when the medium was replaced weekly with conditioned medium they remained stable for as long as 98 days. If 1 percent serum was used for replacement, the cells did slightly less well and the density of the population declined slowly to 4×10^4 cells/cm^2. Figure 2.1 shows what happens if these cells, in which the medium (either 1 percent serum or conditioned medium) is replaced weekly, are continuously labeled with [^{3}H]-thymidine. The cumulative labeling index increases until day 50 to ~70 percent, then increases very slowly from day 50 to day 98 to ~78 percent. It seems therefore that, in these quiescent populations, most of the cells turn over very slowly. This is confirmed by the fact that the daily (24 hour) labeling index under these conditions, was constantly below 5 percent. That

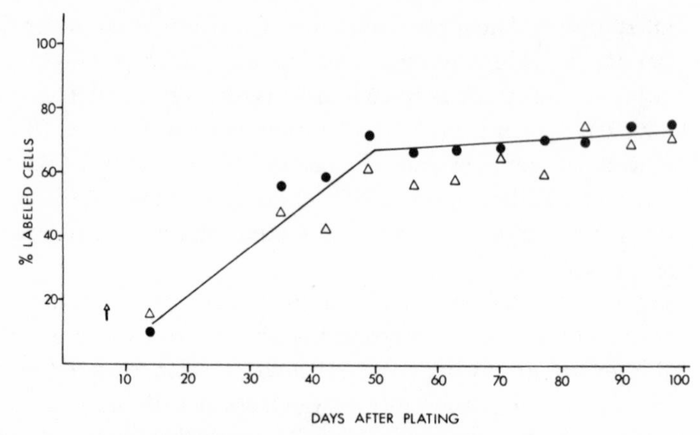

Fig. 2.1 Cumulative labeling index of quiescent WI-38 cells continuously exposed to [^{3}H]-thymidine. The cells were plated and allowed to grow for 7 days in 10 percent serum, by which time the cultures were confluent. At that time (↑) the cells were placed in either conditioned medium (closed circles) or in 1 percent serum (open triangles) and [^{3}H]-thymidine was added to the cultures. The media (always containing [^{3}H]-thymidine) were changed weekly. The cells were fixed at the times indicated on the abscissa. The ordinate gives the fraction of labeled cells in the populations.

the cells were viable and healthy is shown by the results in Table 2.1, clearly indicating that, after addition of serum, even cells kept quiescent for 90 days were capable of re-entering the cell cycle.

In Vivo

In the adult animal, the liver is extremely active in all kinds of metabolic interactions, but, as far as cell division is concerned, it is quiescent. If metabolism and reproduction are the two fundamental characteristics of living cells, the liver is the standard-bearer of metabolism. A single injection of [^{3}H]-thymidine labels only a few cells, whether they are hepatocytes or other cell types, usually <0.2 percent. However, when part of the liver is surgically removed, the remaining liver cells, after a lag period of about 18 hours, enter S phase and eventually divide, restoring the liver to approximately its previous size.

Several studies have been published on the kinetics of entry into S and mitosis of regenerating liver cells after partial hepatectomy.

Table 2.1. Response of quiescent WI-38 cells to a proliferative stimulus (serum).[a]

No. of days after initial plating	% labeled cells after serum stimulation (labeling period in hr)			
	0–24	24–48	48–72	72–96
	CELLS KEPT IN CONDITIONED MEDIUM			
14	31.1	36.0	3.2	2.3
28	17.6	60.3	15.3	16.5
70	34.6	83.3	32.0	3.6
91	15.0	69.5	39.1	10.9
	CELLS KEPT IN 1% SERUM			
14	53.2	57.0	4.6	1.4
28	39.4	82.9	39.0	20.5
49	51.5	70.8	20.7	2.4
91	29.8	80.8	50.8	15.6

a. Cells were initially plated at a density of 2×10^4 cells/cm^2. The medium was changed weekly, and the cultures were confluent by the seventh–eighth day after plating. The cells were then stimulated with 10% serum on the day (after the initial plating) indicated in the first column. Cells were labeled with[^{3}H]-thymidine for 24-hour periods at various intervals after serum stimulation.

Two of the best studies are those by Grisham (1962) and Fabrikant (1968). The main findings are: (1) the lag period is about 18 hours; (2) S phase is 8 hours and G_2 about 3 hours; (3) both parenchymal (hepatocytes) and littoral (Kuppfer) cells participate in the regenerating process; and (4) most cells divide once and a few twice. Liver cells are then an in vivo prototype of G_0 cells—quiescent, nondividing cells that can be induced to re-enter the cycle by an appropriate stimulus.

To round out the picture, the lag period is longer in the liver of old rats, as if the liver cells behaved like WI-38 cells in culture, the length of the pre-replicative phase increasing with the time the cells spend in quiescence. Thus, the peak of DNA synthesis in weanling rats is observed 22 hours after partial hepatectomy, in young adults at 25 hours, and in 1-year-old rats at 32 hours (Bucher 1963). Similar findings have been reported by Adelman et al. (1972) in salivary gland cells that can be stimulated to divide by isoproterenol (see below). In a 2-month-old rat, DNA synthesis reached a peak 18 hours after a single injection of isoproterenol, while in 12-month-old rats the peak was between 36 and 38 hours. Several other examples in which the length of the prereplicative phase is directly proportional to the time spent in quiescence are given in the review by Gelfant (1977). It seems, therefore, that deeper G_0 states exist also in living animals. Indeed, Dethlefsen (1980) lists in his review a number of characteristics that increase or decrease the Q-ness of a cell population, where Q-ness is the degree of quiescence, as opposed to P-ness, which is the proliferating state. Several determinants (cell size, cell density, clonogenicity, etc.) contribute to the Q-ness (or P-ness) of a cell population.

Many other in vivo models of quiescent populations of cells that can be induced to proliferate by an appropriate stimulus have been studied (Table 2.2). Except one, these models have been made obsolete by tissue cultures. However, it is fair to mention them, in expectation of the time, still in the future, when the knowledge we have acquired in vitro will have to be put together in the whole animal. The exception is the lectin-stimulated lymphocyte, and the reason it is an exception is that, in reality, it is an in vitro system. It is listed in

Table 2.2. In vivo models of stimulated DNA synthesis.

Organ or tissue	Stimulus	Reference
Liver	Partial hepatectomy	Grisham (1962)
	Thioacetamide	Reddy, Chiga, and Svoboda (1969)
Kidney	Controlateral nephrectomy	Malt and Lemaitre (1968)
	Folic acid	Baserga, Thatcher, and Marzi (1968)
	Mercuric chloride necrosis	Cuppage and Tate (1967)
	Lead acetate	Choie and Richter (1974)
Salivary glands	Isoproterenol	Barka (1965)
Uterine epithelium	Estrogens	Epifanova (1966)
Mammary glands	Lactation	Stellwagen and Cole (1969)
Spleen	Erythropoietin	Hodgson (1967)
Skin	Wounding	Block, Seiter, and Ohlert (1963)
Lymphocytes	Lectins	Decker and Marchalonis (1978)

Table 2.2 because the G_0 lymphocytes are usually taken directly from the blood of man or other animals. But, after that, everything happens in a test tube. We will meet these lymphocytes several times in this book.

TERMINALLY DIFFERENTIATED CELLS

We have discussed thus far cells that divide continuously and cells that ordinarily do not divide but can do so if an appropriate stimulus is applied. In populations of cells, there is a third category—cells that have ceased dividing and cannot be recalled into the cell cycle. We call these terminally differentiated cells. As an illustration, we shall take the lining epithelium of the mucosa of the small intestine of mammals (Fig. 2.2). The mucosa of the small intestine is lined by a single layer of columnar epithelium, in which three zones can be distinguished: the crypts, the lower part of the villi, and the upper part of the villi. An excellent description of cell proliferation in the lining epithelium of the small intestine of mice has been given by Fry et al. (1963), and our discussion leans heavily on their paper.

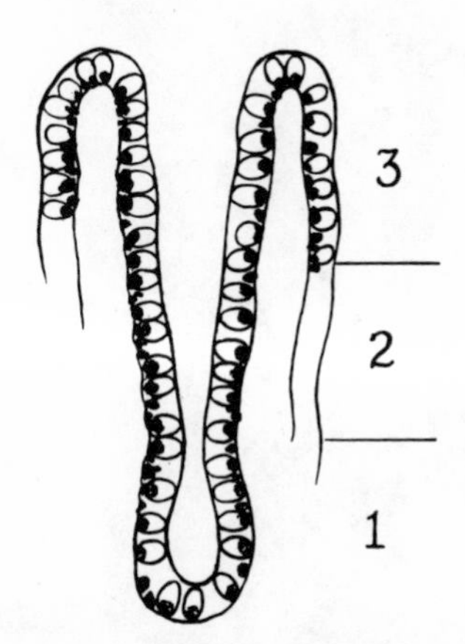

Fig. 2.2 Schematic representation of the lining epithelium of the mucosa of the small intestine. The epithelium consists of a single layer of cells and can be divided roughly into three zones. Zone 1, the crypt, is where all cell divisions occur. Some postmitotic cells then glide along the surface of the villi (zone 2), pushing the preexisting epithelial cells upward. Eventually they reach the tips of the villi (zone 3), from where they desquamate into the lumen of the small intestine. Under physiological conditions, no cell divisions occur in zones 2 and 3.

DNA synthesis and mitoses occur only in the crypts (Fig. 2.3), whose length of about 130 μm is reasonably constant in male and female mice, regardless of age and segment of the small intestine (duodenum, jejunum, or ileum). There are, however, differences in the length of the villi, from ~300 μm in the ileum to ~500 μm in the duodenum, the length also increasing slightly with age. If a mouse is given a single injection of [^{3}H]-thymidine, only epithelial cells in the crypts are labeled in the first few hours after injection. As already mentioned, thymidine is quickly metabolized in animals, so that only cells that were in DNA synthesis at the time of injection and the subsequent 30–45 minutes become labeled. The labeled crypt cells' DNA being now radioactive, their progress from the crypts to the tip of the villi can be followed by killing animals at various intervals after [^{3}H]-thymidine (Fig. 2.4). The transit time from the crypt to the

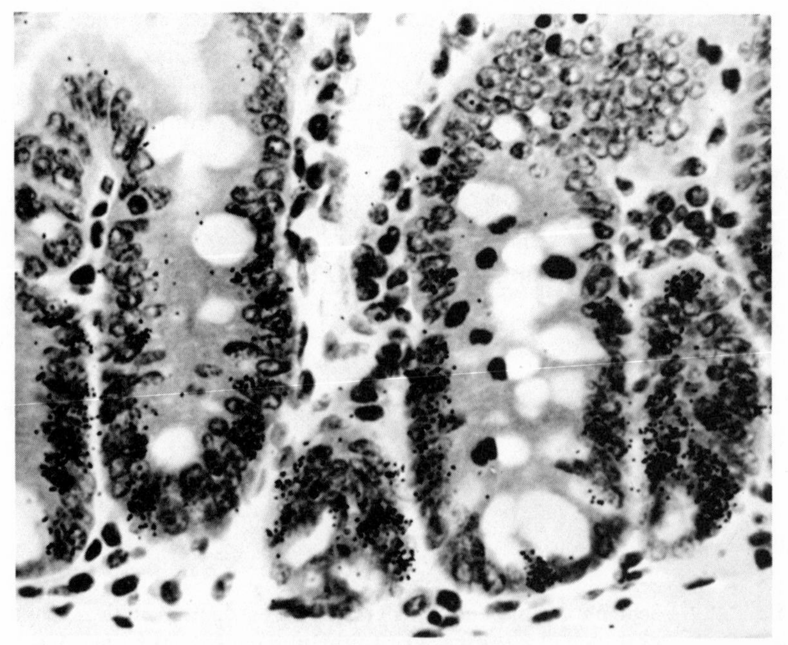

Fig. 2.3 Autoradiograph of the crypt zone of the mucosa of mouse small intestine. The tissues were fixed 30 minutes after a single injection of [^{3}H]-thymidine. Note that the label is limited to the crypt cells. The epithelial cells at the neck of the crypts are not labeled (this is especially clear at the left of the figure). The middle crypt has several mitoses; notice that mitotic cells protrude into the lumen.

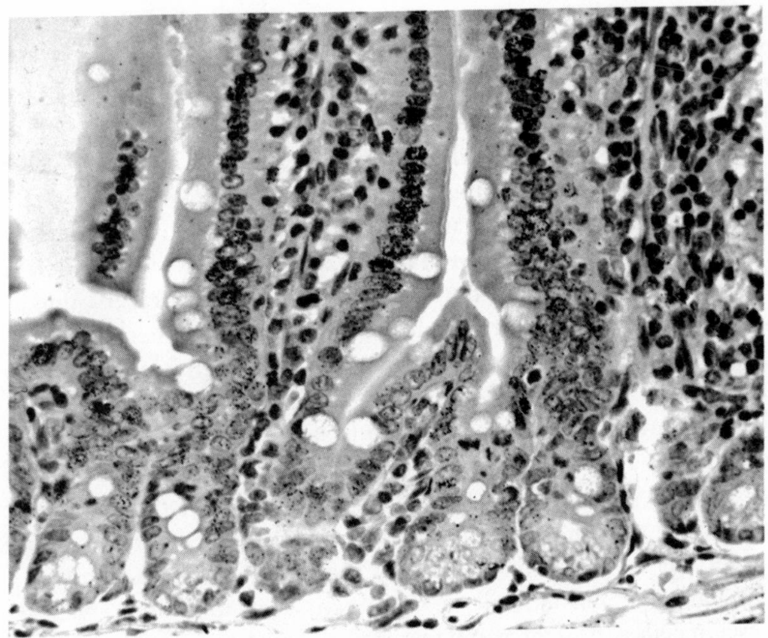

Fig. 2.4 Autoradiograph of the crypts and villi of the mucosa of mouse small intestine. This animal was killed 24 hours after a single injection of [^{3}H]-thymidine. Notice that the epithelial cells lining the villi are labeled. They are the progeny of cells in the crypts that were synthesizing DNA at the time [^{3}H]-thymidine was injected. The label is actually stronger in the epithelial cells of the villi than in those of the crypts. This is because the cells remaining in the crypts have kept dividing, thus diluting the radioactive label.

villus tip is, on the average, about 40 hours. When the cells reach the tip of the villi, they fall into the lumen of the small bowel, where they disintegrate.

The cells at the tip of the villi are terminally differentiated cells, incapable of cell division. There are many cells in animals that can be considered terminally differentiated cells, for instance, polymorphonuclear leukocytes, keratinizing cells of the epidermis and of some mucosae, and, of course, red blood cells, which in mammals are anucleated. In birds, erythrocytes are nucleated, but they are still terminally differentiated. By the way, the presence of a nucleus in chicken red blood cells makes these cells a valuable tool for studies in cell differentiation.

Terminally differentiated cells occur also in tissue cultures. The prototype is the normal human diploid fibroblast, for which the reader is referred to the review by Hayflick (1977). As stated by Hayflick, the spontaneous occurrence of a cell line is a rare event in the cultivation of animal cells. The major exception are mouse cells, which almost always spontaneously transform from a cell strain to a cell line with the concomitant acquisition of the ability to multiply indefinitely in vitro. In 1961 Hayflick and Moorhead, working with cultured normal human fibroblasts, found that these cells underwent a finite number of population doublings and then died. The number of population doublings attained by a given strain of human diploid fibroblasts is inversely proportional to the age of the donor, a finding that has suggested a cellular basis for biological aging (Hayflick 1977). For instance, fibroblasts from fetal lung can undergo as many as 60 population doublings, while fibroblasts from the lung of an 80-year-old donor died after only 18 population doublings. More important for our topic is the fact that, after a certain number of population doublings, all of these cells stop dividing. These senescent cells can be refed and restimulated with serum several times, but they will not re-enter the proliferation cycle. They remain alive in a nonproliferating state for a few weeks and then they degenerate and die.

These cells are terminally differentiated as populations of cells, but Martin et al. (1974), in a seminal experiment, have shown that terminal differentiation occurs even in clones of human skin fibroblasts. In each clone, there is continual selection of stem cells that segregate daughter cells of varying growth potential, including a class of cells which may be regarded as terminally differentiating. In this system, cell senescence seems to be a stochastic process, perhaps not surprising, since the greying of our hair and beard also has all the appearances of a stochastic process.

In addition to normal human fibroblasts, there are now available several other in vitro systems in which differentiation can be induced by appropriate manipulations. These systems are very useful for investigating changes in gene expression during differentiation. The most commonly used include: (1) L6E9 rat myoblasts that grow vigorously in culture but can be induced to fuse into nonproliferating myocytes by incubation in "differentiation medium" (Nadal-Ginard 1978). Markers of differentiation (for instance, creatine phosphokinase and myosin heavy chains) can be detected in the fused cells; (2) human epidermal keratinocytes that can be grown on a feeder layer of irradiated 3T3 cells. These keratinocytes form colonies that undergo terminal differentiation, as characterized by an increase in size and the development of an envelope resistant to detergents and reducing agents. Addition of epidermal growth factor delays the terminal differentiation of these keratinocytes (Rheinwald and Green 1977); (3) human promyelocytic leukemia cells, HL60, that can be induced to differentiate by the addition of phorbol esters. Induction of HL60 cells causes them to differentiate terminally into cells with all the characteristics of macrophages (Rovera, Santoli, and Damsky 1979); (4) liver or placental cells transformed by a tsA mutant of SV40. At the permissive temperature (33°) these cells grow like transformed cells. However, when switched to the restrictive temperature (40°) that inactivates the A gene function of SV40, growth slows down and, for instance, the liver cells express markers of differentiation, such as maximal synthesis of α-fetoprotein, albumin, and transferrin (Chou and Schlegel-Haueter 1981); and finally (5) teratocarcinoma cells, where the stem cells can be induced to differentiate by retinoic acid (see review by Strickland 1981). Several other systems are available and these have been listed in a careful review by Abraham and Rovera (1980). These authors also discuss the seemingly contradictory effects of phorbol esters, which inhibit differentiation in some systems (like mouse erythroleukemia cells) and induce it in others (see, above, the HL60 cells).

Chapter 3

Populations of Cells

We have seen that most populations of cells consist of a mixture of three different subpopulations—continuously cycling cells, G_0 cells, and terminally differentiated cells—and that immediately after fertilization of the egg, all cells divide (Graham and Morgan 1966). For instance, in *Xenopus laevis* embryo, at stage 7, all cells are labeled by a 10-minute exposure to [^{3}H]-thymidine. The division of all cells, combined with a very short cell cycle (about 14 minutes), results in extremely rapid growth—from 50 cells at stage 6 (3.5 hours after fertilization) to over 5,000 cells at stage 8 (5 hours after fertilization). However, between cleavage stages 11 and 13, when the total number of cells in the embryo is past 10^4, some cells stop dividing and form a population of nondividing cells (see also Newport and Kirschner 1982). Sometime during fetal development (when exactly is not known) some cells begin to die, others keep cycling, and still others remain alive but do not divide. From this moment, all populations of cells, whether normal or abnormal, can be said to consist of the three distinct subpopulations mentioned above. This is also true of tumors, as reviewed by Dethlefsen, Bauer, and Riley (1980), who distinguish three populations of cells in solid tumors: Q cells, P cells, and D cells, corresponding to our G_0, cycling, and dying cells, respectively. As we shall see later, this is a very important consideration in our understanding of how populations of cells, including tumor cells, grow in number. The fraction of cells in the population that are cycling is called the "growth fraction," a concept originally introduced by Mendelsohn (1962). To quote from his paper: "Growth fraction is defined as the proportion of proliferating or growing cells in a population. For present purposes, a proliferating or growing or dividing cell is defined as a cell actively involved in the mitotic cycle." Mendelsohn devised an autoradiographic method to estimate the growth fraction in a mouse breast cancer and, although some refinements were introduced in later years, his methodology remains basically the one of choice.

An important thing to remember at this point is that cells can move freely from one compartment to another, that is, they can enter G_0 or re-enter the cell cycle depending on the environmental conditions. The only forbidden transition is a reversal of terminally differentiated cells into the proliferating pool or into G_0. It is true that nuclei of terminally differentiated cells can be induced to enter S phase by fusion to rapidly growing cells (see chapter 7), but this is a special case that should not invalidate our statement on forbidden transitions.

In cultured cells, especially in monolayers, one can observe a similar mixture of subpopulations. Newly plated cells have a lag phase, during which very little growth occurs. For instance, with human diploid fibroblasts, Norrby (1970) could not detect mitoses until 8 hours after subcultivation, and mitotic activity remained low up to 24 hours. This is followed by an exponential growth phase, during which nearly all cells are dividing, which leads to a rapid increase in cell number. A third phase follows, during which the number of cells still increases but at a continuously decreasing rate, and finally the cell population reaches a plateau phase, when the number of cells no longer increases. Since, during the plateau phase, a few cells can still be seen cycling, some cells must be dying, to keep the number unchanged; indeed, Norrby (1970) has observed cell death in healthy cultures of normal Syrian hamster cells.

This distinction between cycling, dying, and G_0 cells is probably already a simplified view of cell populations. It is worthwhile at this point to reproduce in almost its entirety a table from the paper by Darzynkiewicz, Traganos, and Melamed (1980) which subdivides each phase of the cell cycle into subcompartments that can be distinguished by multiparameter flow cytometry (Table 3.1). Readers who don't believe in G_1 and G_0 can skip this table. I think it is useful, provided one does not try to apply it rigidly to all populations of cells. The most remarkable compartment, to me, is S_Q, a population of presumably live cells arrested in S. We do not have, however, any evidence that S_Q cells are actually viable, i.e., capable of proliferating. On the other hand, G_{2Q} makes sense. These are essentially tetraploid cells: their low amount of RNA could explain why there is a pre-replicative phase even in regenerating liver after partial hepatectomy, since most of the hepatocytes have a 4n amount of DNA. Indeed, the reader is referred to an excellent review by Gelfant (1977), who lists four major categories of cells: cycling cells, noncycling G_1-blocked cells, noncycling G_2-blocked cells, and noncycling G_0 cells. Cells arrested in G_2 are obviously recognizable by their DNA content, but Gelfant also distinguished G_0 cells from G_1-blocked cells on the basis of the time required to enter S after the application of a stimulus (see also chapter 5).

Although all populations of cells can be reduced to the above mentioned subpopulations, there are two other types of cells that should be described at this point: stem cells and transformed cells. Before turning to a consideration of these cells, I would like to remind the reader of the heterogeneity of seemingly homogeneous populations of cells. Thus, in cultures of human diploid fibroblasts morphologically homogeneous, certain growth factors stimulate only a specific subpopulation which can then be selected by appropriate methods. For instance, from a morphologically homogeneous population of human gingival fibroblasts, Bordin, Page, and Narayanan (1984) were able to select a subpopulation that had a high affinity for the first component of complement, C1q, to which it responded with cell proliferation, and that produced unusually large amounts of collagen. The world of cell cultures is less uniform than it may appear at first sight.

STEM CELLS

The concept of stem cells originated with the pioneer work of McCulloch and co-workers. The assay for stem cells was first described by Till and McCulloch (1961). A suspension of bone marrow cells was injected intravenously into mice that had received a lethal dose of X-irradiation. With the endogenous bone marrow cells destroyed by radiation, the injected cells formed colonies in the spleen. There was a linear relationship between colony formation and the number of bone marrow cells injected, and the efficiency was 1 colony per 10^4 cells. Cytological studies indicated that the spleen colonies were clones (i.e., they originated from single cells) and contained a mixture of erythroid and granulocytic cells (Becker, McCulloch, and Till 1963). Besides precursors of one or more of the haemic lines, the spleen colonies also contained cells that formed colonies when transferred to a second irradiated recipient mouse. To quote from Potten, Schofield, and Lajtha (1979): "The cells from which the colonies derive exhibit pluripotency and self-renewal potential, i.e., satisfy the criteria for hemopoietic stem cells." More recently, other criteria have been proposed for the identification of stem cells. In a seminal paper Korsmeyer et al. (1981), working with leukemic cells and using the most sophisticated methods of recombinant DNA technology, have shown that in stem cells there is no rearrangement of immunoglobulin genes, while gene rearrangement is already present in pre-B cells.

Stem cells are not the only cells capable of division. Some more differentiated precursor cells are also capable of cell division so that growth in number depends on two categories of cycling cells: stem cells and amplifying transit cells. However, the latter cells have a limited number of divisions. We can now define the stem cells, again quoting the excellent review of Potten, Schofield, and Lajtha (1979): "The stem cell represents a minority class of slowly cycling, self-maintaining cells that occupy some specific positions within each tissue, from which are derived a second larger population of rapidly dividing transit cells with limited or no self-replicative ability." The present discussion is based on this review and on a subsequent paper by Potten and Lajtha (1982).

Although the bone marrow has been the best model for stem cells, Potten, Schofield, and Lajtha (1979) have made a good case for the existence of stem cells in other tissues (intestinal epithelium, tongue

Table 3.1. Main features of the respective phases of the cell cycle that may be distinguished by multiparameter flow cytometric cell analysis.[a]

Phase	DNA (C)	RNA	Next phase transition	Separation from other phases	Examples
G_{1Q}	2	Very low	G_{1A}; slow; exponential	No overlap	Peripheral blood lymphocytes; confluent 3T3 cultures
G_{1A}	2	Low	G_{1B}; fast; exponential	Continuity G_{1B}	All cycling populations; A/B ratio changes with change of growth rate
G_{1B}	2	High	S; linear (?)	Continuity with G_{1A} and S	
G_{1D}	2	Varies depending on cell type	Cell death	Often full separation	Granulocytes, differentiated FL cells, epidermal, plasma or mast cells
S	$2 < C < 4$	High, often increasing	G_2	Continuity with G_{1B} and G_2	All cycling populations

S_Q	$2 < C < 4$	Low	S	No overlap with S	Some leukemias (i.e., chronic myloid leukemia-blastic crisis) and solid tumors
G_2	4	Nearly twice G_{1A}	M	Continuity with S	All cycling populations
G_{2Q}	4	Low	G_2	No overlap with G_2	Some leukemias and solid tumors
M	4	Twice G_{1A}	G_{1A}	No overlap with G_2	All cycling populations
			CELLS IN TRANSITION		
G_{1T}	2			Between G_{1Q} and G_{1A}	Phytohemagglutinin-lymphocytes; serum deprived 3T3 cultures
S_T	$2 < C < 4$			Between S_Q and S	Chronic myeloid leukemia-blastic crisis; cells in short-term culture
G_{2T}	4			Between G_{2Q} and G_2	

a. RNA estimates relate to the same cell type in other phases of the cycle. From Darzynkiewicz, Traganos, and Melamed (1980).

mucosa, epidermis, testis) where the micromilieu, which they call "niche," could determine proliferation and differentiation probabilities.

The two cardinal properties of stem cells are pluripotentiality and self-maintaining capacity. They are usually also clonogenic, but as pointed out by Potten and Lajtha (1982), colony-forming cells in the mouse hemopoietic system (CFU-S) *are* stem cells, but not all stem cells are CFU-S. In the crypts of the small intestine, the opposite is observed: there are many more clonogenic cells than stem cells, a ratio of about 60 to about 16, and some of the stem cells are not clonogenic. Stem cells are therefore pluripotent and self-maintaining, whereas transit cells are not. The majority of the stem cells are slowly cycling. For instance, in the normal bone marrow, fewer than 1 percent of the cells are stem cells; and of these, only 10 percent are cycling at any given time. According to Potten and Lajtha (1982), the total stem cell population represents only 0.25 percent of the marrow cellularity, but one stem cell per day differentiating can maintain some 4,000 cells in amplifying transit maturation and about

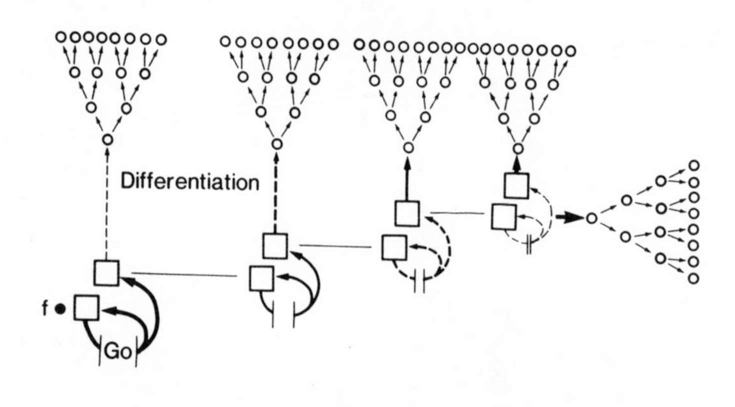

Fig. 3.1 A general scheme for cell replacement and differentiation. The boxes are stem cells and circles transit cells (see text). The stem cells are at a hypothetical focal point, f, with restrictions imposed upon them by the local tissue microenvironment. Only one daughter cell at each division remains at the focal point; the other daughter cell is pushed into the first of a series of shells (b–d) of declining stem cell quality and increased differentiation probability. Note that many stem cells are in G_0. (Reprinted, with permission, from Potten, Schofield, and Lajtha 1979.)

1,000 granulocytes or 200,000 erythrocytes in the peripheral blood. A general scheme for stem cell proliferation and differentiation is shown in Fig. 3.1, where stem cells are pushed into niches of declining stem cell quality or "stemness," while transit cells amplify the population, differentiate, and eventually die. In some tissues, like the crypts of the small intestine, the niche can be identified: here the primary stem cells are located in a circumferential ring of 16–18 cells just above the Paneth cells (Potten and Lajtha 1982).

Stem cells have also been described in tumors, but it is not clear there whether they constitute a population of rapidly dividing cells or whether they are made up, like normal stem cells, of discrete subpopulations of proliferating and nonproliferating cells. A good review of the problem of tumor stem cells can be found in Steel (1977). I will limit myself here to a reminder that stem cells are the key to successful chemotherapy of cancer: tumor stem cells must be destroyed and normal stem cells must be, at least partially, preserved. Transit cells, because they have a limited proliferative potential, can (at least in theory) be ignored.

TRANSFORMED CELLS

The definition of a transformed cell is relatively easy, if one limits oneself to cells in cultures transformed by certain viruses. It becomes more complicated if cells in cultures are transformed by chemical or physical agents, and quite difficult if one wishes to make transformed cells the equivalent of neoplastic cells.

Virally-Transformed Cells in Culture

The prototypes are cells transformed by SV40 or polyoma virus. Ponten (1971) has given a lucid analysis of transformation, and we will follow closely his book in the initial part of this discussion. According to Ponten, transformed cells are characterized by: (a) irregular growth; (b) unrestrained growth; and (c) infinite growth.

Irregular growth transformation is due to lack of contact inhibition of movement, so that cells can move over each other to form randomly arranged multilayers, or, as it is often stated, "they can pile up." Normal cells in culture form monolayers and do not pile up. There are already two problems here. Under certain conditions normal cells *can* form multilayers. For instance, using the perfusion-culture technique, normal human diploid cells can form up to 4–6 layers of cells (Kruse, Whittle, and Miedema 1969). On the other hand, under standard conditions 3T3 cells form monolayers without piling up; yet by other criteria 3T3 cells are transformed.

Unrestrained growth transformation is the loss of contact inhibition of cell proliferation, so that the transformed cells reach a higher saturation density than a normal cell population. All cell lines and cell strains have a characteristic saturation density—a maximum number of cells per unit area of surface supporting cell growth. To quote from Ponten (1971): "Saturation cell density is independent of the size of the inoculum used to start the culture, but differs between different types of cells and is also influenced by the composition of the medium." In general, if one compares a cell line with its virally-transformed counterpart, one can say that transformed cells grow at a higher saturation density, especially at low concentrations of serum. Indeed, we can say that, as a rule, the serum requirements are lower in transformed cells, so that transformed cells can grow in concentrations of serum (or growth factors) that do not allow the growth of normal cells.

Infinite growth transformation is the loss of aging properties, permitting cells to undergo unlimited division in culture (and to establish cell lines). We have seen in chapter 2 that normal human fibroblasts have a limited life-span in culture and cease to proliferate after a certain number of population doublings. Transformed cell lines have an indefinite growth potential—some colleagues say they are "immortalized." Since the oldest cell line around is about 40-years-old I submit that to call these cell lines "immortal" is perhaps premature. However, the term "immortalization of cells" (perhaps as a subconscious expression of our existential condition) has now become so prevalent that no amount of logic will be able to repeal it. Apart from nomenclature, there are cell lines with an indefinite growth potential that are contact-inhibited both for movement and for growth. Indeed, for many of us, the "normal" cells to which transformed cells are compared are often 3T3 cells, which have an indefinite growth potential and are grossly aneuploid.

Clearly then, some other, more stringent criteria are necessary for the definition of a transformed cell line. One feature that is widely accepted as characteristic of a transformed cell line is anchorage independence, or the ability to grow unattached to the supporting glass or plastic surface. This orginated in the observation of Macpherson and Montagnier (1964) that BHK cells transformed by polyoma, or infected with polyoma, were capable of forming colonies in agar, while normal BHK cells did not. Since then, anchorage independence (ability to grow in agar or methylcellulose) has been considered a valid criterion of transformation (see, for instance, Frisque, Rifkin, and Topp 1980; Hiscott and Defendi 1980). Here again, exceptions can be found. Thus, certain lymphoblastoid cell lines — without any detectable viral insert, and presumably normal — have an indefinite growth potential and grow in suspension. Also, certain growth factors, especially a class of growth factors called transforming growth factors (chapter 10), can confer anchorage independence to untransformed cells. Despite the exceptions, anchorage independence is still a valid criterion of transformation. One should remember, though, that the anchorage dependence of normal cells is limited only to a short period of the cell cycle, in late G_1 (Matsuhisa and Mori 1981).

Cells Transformed by Chemical Agents

Cells transformed by chemical carcinogens behave somewhat differently from virally-transformed cells. Cells transformed, for instance, by 3-methylcholanthrene grow at a higher saturation density than their normal counterparts. However, at variance with virally-transformed cells, they can arrest in G_1, like normal cells (Moses et al. 1978). In chemically transformed cells, growth is arrested in G_1 by a different mechanism than in untransformed cells and can be restimulated simply by the addition of nutrients, while normal cells require growth factors (Moses et al. 1978).

Cherington, Smith, and Pardee (1979) studied the requirements for growth factors of transformed and untransformed hamster cell lines growing in serum-free media. Polyoma-transformed BHK-21 cells no longer required the FGF, EGF, and insulin that were required by BHK-21. They retained a requirement for transferrin. Chemically transformed cells decreased their EGF requirements but still needed insulin and transferrin.

Relationship between Transformation in Vitro and Neoplasia in Vivo

The ultimate test of transformation is the ability to induce tumors in appropriate host animals. Syngenic animals or nude mice are often used to test the tumor-inducing ability of transformed cells.

Table 3.2. Criteria of transformation.[a]

Minimal transformation	Intermediate transformation	Full transformation
Indefinite growth potential	Clonal growth in agar or methycellulose	Ability to produce tumors in animals
High saturation density		
Clonal growth on plastic		
Growth in low serum		

a. Transformation should be viewed as a multistep process, in which cells progress from minimal transformation to full transformants capable of inducing tumors when injected into appropriate host animals. Thus, intermediate transformants have all the characteristics of minimal transformants, *plus* the ability to form clones in agar or methylcellulose.

As stated by Ponten (1971); "At first sight, such a procedure would seem to provide a clear-cut answer; however further analysis shows that definite conclusions can be made only with caution." Genetic incompatibility, tumor cell antigens, the number of cells implanted, the site of inoculation, and other factors may affect the tumor-inducing ability of implanted cells. In general, though, we can say that cells capable of inducing tumors in animals have all the other characteristics of transformed cells. We can therefore list in Table 3.2 the criteria for transformation, as discussed in this chapter. Some criteria are relative because they are shared by some untransformed cells. However, when all these criteria are present, one can reasonably call these cells minimal transformants. These minimal transformants can become more transformed. To paraphrase from Orwell, all transformed cells are equal, but some transformed cells are more equal than others—and this super-equality is characterized by the acquisition of anchorage independence. Finally, the fully transformed cell is the equivalent of the cancer cell, with all the attributes of transformation.

The progressive increase in malignancy of certain neoplasms has been described not only in animals but even in man and, more recently, in tissue cultures. Although this is not the place to discuss it, tumor progression is a reality (Shubik, Baserga, and Ritchie 1953) that could possibly be based on the progressive acquisition of the various characteristics of transformation. Again, an in vitro model has emerged in more recent years with the studies of Ts'o and collaborators on carcinogen-treated Syrian hamster embryo cells in culture. When treated with a carcinogen such as benzo (a) pyrene, Syrian hamster embryo cells can undergo malignant transformation, but the acquisition of phenotypic markers often associated with neoplasia is progressive (Barrett and Ts'o 1978). Thus, altered morphology and loss of contact inhibition precede the ability to grow in soft agar (Nakano et al. 1982), which in turn correlates with tumorigenicity. The evidence for the progressive nature of neoplastic transformation raises a number of questions that have now been addressed at the molecular level and will be discussed in chapter 13.

Chapter 4

Tissue Growth

Since tissues and organs consist of populations of cells held together by varying amounts of intercellular substance, tissue growth can be due to: (1) an increase in the number of cells; (2) an increase in size of the cells; (3) an increase in the amount of intercellular substance. However, the intercellular substance of a tissue is usually a secreted product of the cell (for instance, collagen), an extracellular extension, so to speak, of the cytoplasm. We can therefore consider an increase in intercellular substance as a variation of an increase in cell size, and thereby reduce tissue growth to two mechanisms, growth in size and growth in number of cells. These two mechanisms of growth are so basic that they can be found even in unicellular organisms. Harris, in his delightful book *Nucleus and Cytoplasm* (1968), has described in detail the unicellular alga *Acetabularia mediterranea*, which grows from a microscopic zygote into a plant-like structure 3.5 cm long that contains a single nucleus located in the tip of one of the rhizoids at the base of its stalk. *Acetabularia*, that is, grows exclusively by increasing its size. A single bacterium, on the other hand, will produce a colony visible to the naked eye through an increase in the number of bacterial cells. As an illustration of these two mechanisms in higher organisms, we may take the deer's antlers and the cock's comb. The extraordinarily rapid growth of the deer's antlers in spring is due to a continuous proliferation of progenitor cells at the base of the antlers. The cock's comb instead, grows by increasing the size of cells, especially the amount of intercellular substance, which is largely hyaluronic acid and its derivatives.

GROWTH IN SIZE

Cells can grow in size in three ways: (1) during the cell cycle; (2) during development; and (3) under pathological conditions.

Growth in Size during the Cell Cycle

When a cell goes from one mitosis to the next one it doubles its size. This is intuitive, since if dividing cells were not to double their size from G_1 to M, they would become progressively smaller at each division, and eventually vanish. During balanced growth under ordinary conditions, to quote from Fraser and Nurse (1978): "The two daughter cells produced at each division are identical to the parent at the same time in the preceding cycle: this requires that all cell components are doubled during the course of each cell cycle." In studying cell proliferation, most of us are so preoccupied with cell DNA synthesis (probably because it is so easy to measure) that we have often forgotten that growth in cellular size is just as critical for cell division as DNA replication. We measure growth signals by measuring the incorporation of [^{3}H]-thymidine into DNA, but we often ignore the other question: what are the signals that tell the cell to double its other cellular components? We shall see later how important this question is, since growth in size and cell DNA replication can be dissociated (see chapter 9). We are also postponing until that chapter the controversial question of whether or not cell size controls cell DNA replication. For the moment, let us give credit to

Mitchison (1971), who, in studying the yeast cell cycle, proposed several years ago that there are not one but two cell cycles: a growth cycle, that occupies the whole cell cycle and during which the cell progressively grows in size, and a nuclear cycle, that occupies only a discrete portion of the cell cycle and during which cell DNA is replicated. This concept is represented diagrammatically in Fig. 4.1.

To measure growth in size of a cell, one can directly measure cell mass. However, proteins and nucleic acids are also good indicators of cell size (Cohen and Studzinski 1967), first because they constitute 50 percent of the dry weight of cells and second because on them largely depend the amounts of other components such as sugars and lipids. In careful studies, using microspectrophotometry and microinterferometry of single cells in culture, Killander and Zetterberg (1965) showed that the dry mass content and the RNA content of mouse fibroblasts grew progressively with cell age, i.e., from mitosis to G_2 (Fig. 4.2). These authors also analysed the distribution of these substances at mitosis, by measuring DNA, RNA, and mass in sister pairs. They found that the error of DNA distribution during mitosis was very low, in the same order of magnitude as the error of measurement, while the error of distribution of RNA and particularly of mass was higher than the error of measurement. Two more things

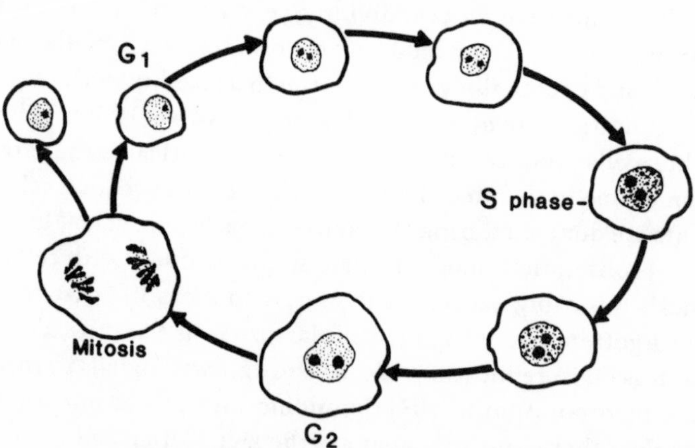

Fig. 4.1 Growth cycle and nuclear cycle in animal cells. After mitosis, a cycling cell immediately begins the growth cycle and progressively increases in size until the next mitosis. The DNA synthesis phase, represented here by the medium-sized grains on the fourth and fifth cells after mitosis, occupies only a discrete portion of the cell cycle.

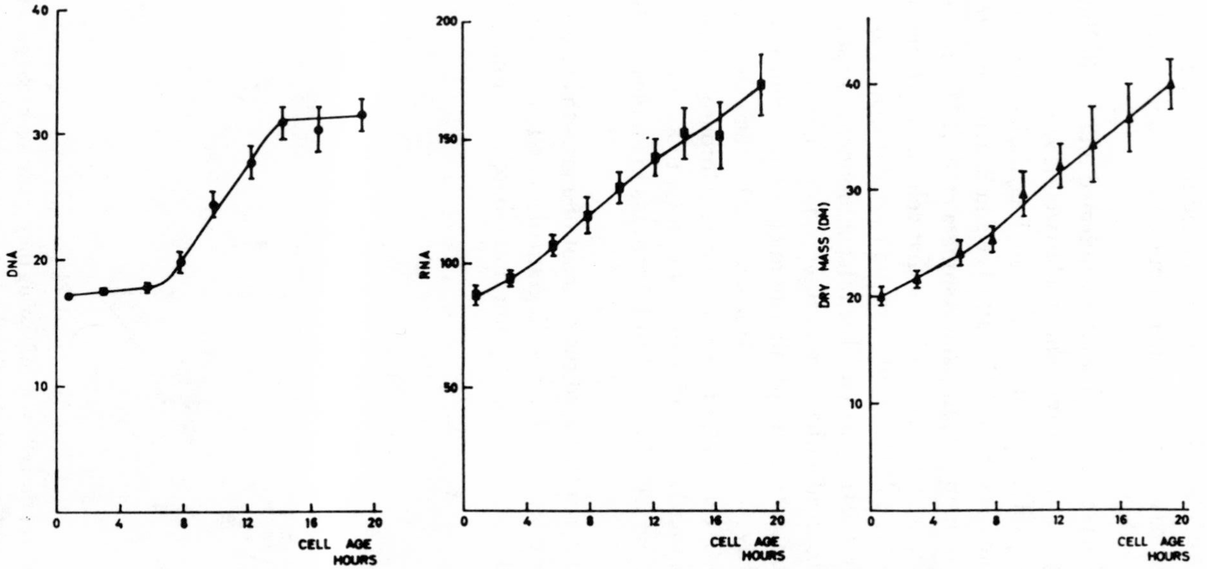

Fig. 4.2 Dry mass and amounts of DNA and RNA in cells progressing through the cell cycle. Abscissa is time after mitosis. Left panel = DNA amount per cell; middle panel = RNA amount per cell; right panel = dry mass. Notice that RNA and dry mass increase almost linearly throughout the cell cycle, while the increase in DNA content is discontinuous. (Reprinted, with permission, from Killander and Zetterberg 1965.)

should be noted: cellular dry mass is directly proportional to protein content (Gaub, Auer, and Zetterberg 1975), and RNA content doubles during interphase in the cytoplasm but remains unchanged in the nucleus (Zetterberg 1966).

These studies have been repeatedly confirmed and I will limit myself to citing only a few references: Hartwell (1978) and Skog, Eliasson, and Eliasson (1979) for cell size doubling; the latter authors and also Baxter and Stanners (1978) and Lee and Engelhardt (1977) for doubling of protein content; and Ashihara et al. (1978) for the doubling of RNA content. The point is that cell mass, amount of protein, or amount of RNA are all good indicators of growth in cell size. Incidentally, the truism that all cellular components must double can be also observed with low molecular weight compounds. For instance, the concentrations of all ribonucleoside triphosphates double from early G_1 to S: in CHO cells, UTP increases from 1.5 nmoles per 10^6 cells in G_1 to 3.1 in S phase (Hordern and Henderson 1982). Similarly ATP goes from 4.0 in G_1 to 8.7 in S, GTP from 0.6 to 1.6, and CPT from 0.7 to 1.7 (always in nmoles per 10^6 cells).

Now that we have introduced the concept of a growth cycle and also of G_0 cells, we can complete the diagram of Fig. 1.1, as shown in Fig. 4.3. The cell enters G_0 from a point immediately after mitosis,

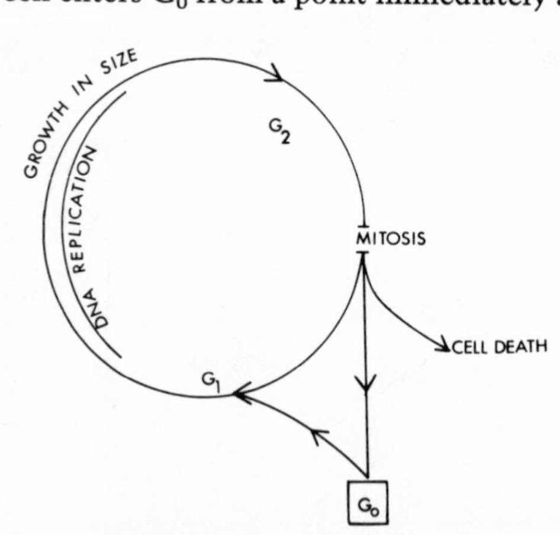

Fig. 4.3 Diagram of the cell cycle indicating the separation of the growth cycle and the nuclear cycle. We have also added the G_0 state, while cells leaving the cell cycle (i.e., terminally differentiated cells) are represented by cell death since, in terms of cell proliferation, these cells are destined to die without dividing again. This division of cell populations into cycling cells, G_0 cells, and dying cells will be useful when considering the growth of tumors.

but re-enters the cell cycle in mid-G_1. This reflects some uncertainty on what actually happens, although evidence is growing that mid-G_1 may be the point of both exit from and re-entry into the cell cycle (see chapter 11).

Growth in Size during Development

Growth in size during the cell cycle is reversible, the cells halving their size at mitosis. However, during development, many cells grow in size, and the increase in size becomes a permanent feature of the cell. For instance, myocardial fibers in man increase in diameter from 7 μm at birth to 14 μm in the adult. An increase in protein/DNA ratio is also a good indicator of an increase in cell size. The protein/DNA ratios increase from newborn to adult man, in skeletal muscle from 120 to 206, in liver from 29 to 73, and in kidneys from 23 to 60 (Widdowson 1981). A good illustration of growth during development is given by the postnatal growth of rat liver, as summarized in Table 4.1 from the paper of Fukuda and Sibatani (1953). There is an 11-fold increase in the number of cells, a doubling in DNA content per cell (diploid liver cells become tetraploid), while the amount of protein per cell increases 5 fold, and cell mass almost 4 fold. While the increase in cell number predominates, it is also clear that there is an increase in cell size.

Growth in Size under Pathological Conditions

A number of human diseases, as well as experimental manipulations in laboratory animals and cell cultures, cause a marked increase in the size of cells, way above their usual size. Borrowing a term from classical pathology, we call an increase in cell size "hyper-

Table 4.1. Postnatal growth of rat liver.

Age (in days)	Liver weight (in g.)	Cell number ($\times 10^{-6}$)	DNA per cell (av. in pg)	Protein per cell (av. in pg)	Cell mass (ng)
10	0.30	168	5.9	29.3	1.79
21	0.98	445	5.1	45.6	2.20
41	5.7	1060	11.1	103.0	5.36
80	8.1	1270	11.4	154	6.37
182	12.0	1790	11.4	155	6.70

Modified from Fukuda and Sibatani (1953).

trophy." Although a modest increase in cell number also usually occurs during hypertrophy, it is an increase in cell size that predominates.

An example of hypertrophy is what happens in the myocardial fibers of patients with hypertension. The heart markedly increases in weight, the diameter of the myocardial fibers increases several fold, and, interestingly, there is a concomitant increase in the amount of DNA per nucleus. We have mentioned before that rat liver cells have a diploid content of DNA at birth (~6 pg per cell) but become tetraploid during postnatal development. A similar situation occurs in the human heart. In children below 3 years of age, 85 percent of the myocardial nuclei are diploid and 15 percent are tetraploid (Fischer et al. 1970). In adult hearts of normal weight, 95 percent of the myocardial nuclei are tetraploid. However, in hypertrophy of the heart, not only the fibers increase in diameter (100 μm), but most of the nuclei become octaploid, and some may even have 32n amounts of DNA (Sandritter and Scomazzoni 1964). Another good example of hypertrophy in man is the megaloblast, a markedly enlarged precursor of red blood cells, also with an increased amount of DNA, that appears in patients with a deficiency of vitamin B12. Hypertrophy can also occur in nonpathological conditions, as for instance in athletes' response to training. The most striking example is the hypertrophy of the heart and skeletal muscles that occurs in runners, whether they are of the human or equine variety.

A good model of experimental hypertrophy is the isoproterenol-stimulated salivary gland of mice or rats. Although the first administration of isoproterenol stimulates cell proliferation in acinar cells, repeated injections result in marked cellular hypertrophy with very little cell division. Thus, after 10 days of chronic treatment, mitoses are no longer detectable, while the salivary gland's weight continues to increase and the amount of DNA per nucleus goes from 8 to 29 pg (Novi and Baserga 1971). Hypertrophy, rather than hyperplasia, also characterizes the increase in mass of the arteries in spontaneously hypertensive rats (Owens, Rabinovitch, and Schwartz 1981).

GROWTH IN NUMBER

It does not take a profound observer to realize that the most important mechanism in the growth of tissues is an increase in cell number, since all animals derive from a single fertilized egg cell. The adult rat liver alone has almost 2×10^9 cells, though the liver is much more cellular than the animal as a whole, which contains bones, tendons, cartilage and other tissues that have a lot of intercellular substance and few nuclei. It is possible to calculate, approximately, the number of cells in an animal by measuring the total amount of DNA. For instance, a mouse 25 g in weight contains a total of 20 mg of DNA. Since diploid somatic cells (in mammals) have 6×10^{-12} g of DNA, one can calculate that there are, in mouse, $\sim 3 \times 10^9$ cells. By extrapolation, a 70 kg man should have in the order of 10^{13} cells. A slightly different estimate — 3.5×10^{13} cells — is obtained if one counts the number of cells in 1 g of tissue, which has roughly 5×10^8 cells. Incidentally, as one can see from Table 4.1, the mass of a diploid liver cell in the rat is 2×10^{-9} g, which again would give 5×10^8 cells/g. These figures are very approximate and should serve only as a guide.

Whether in cell cultures or in tissues, growth results when the number of cells that are produced per unit time exceeds the number of cells that die in the same period of time. In the adult animal, even in rapidly proliferating tissues such as bone marrow and the lining epithelium of the gastrointestinal tract, the number of cells produced equals the number of cells that die, and these tissues are said to be in a steady state. In old animals the number of cells that die sometime exceeds the number of cells that are produced. This results in the atrophy of some tissues. In humans this is especially evident in the brain, where, with age, there is a loss of 150 g (average) or the equivalent of 7.5×10^{10} cells. Although some of these lost brain cells certainly contain valuable information, let us console ourselves by reflecting that many of them contained misinformation, that is thus properly consigned to oblivion.

Normal growth extends from fertilization to adulthood. Characteristically, the specific growth rate of organisms declines exponentially with age, and varies from organ to organ. As an example, the human body weight increases constantly until adolescence, after which the rate of increase slows. The weight of the human brain, however, increases quickly from birth to the fourth year of age (from 350 g to 1300 g) and then very slowly until the eighteenth year of age (1450 g).

There are two forms of growth in cell number that should be considered at this point: hyperplasia and neoplasia. Hyperplasia is a localized increase in cell number. It can occur under physiological conditions, for instance, the enlargement of the glandular epithelium of the mammary gland during pregnancy and lactation. When lactation is discontinued, the mammary gland involutes and returns to its normal size. In pathologic hyperplasia, the localized increase in cell number is due to abnormal stimuli and persists at least as long as the stimulus to hyperplasia is present. That makes hyperplasia different from neoplasia or cancer, for which the best definition remains that of Willis (1952): "A neoplasm is an abnormal mass of tissue, the growth of which exceeds and is uncoordinated with that of the normal tissues and persists in the same excessive manner after cessation of the stimuli which evoked the change." We will discuss later how tumors grow.

Cells in cultures also increase in number. The growth curves that are generated by counting the number of cells can be divided into three phases: a lag phase, usually in the first 24 hours after plating, an exponential growth phase, and a stationary phase. In the last phase, there is no net increase in cell number. The doubling time in the exponential growth phase gives a measure of the rapidity with which the cells divide, while the maximum number attained gives us the saturation density of the cell line.

It seems almost obvious, but I must warn the reader that cell DNA synthesis is not the same as cell proliferation. Too often we read papers stating that a certain substance causes cell proliferation, only to find out that cell proliferation was determined by autoradiography of cells labeled with [^{3}H]-thymidine. Cell proliferation means that cells are dividing, and the only way to determine the number of cells in a culture dish is to count them. Another important distinction is between G_0 cells and cells in stationary cultures. When certain cells reach saturation density, most of them enter the G_0 state and there is no increase in cell number and no (or very little) cell proliferation. However, under certain nutritional conditions and particularly with transformed cells, one can have stationary cell cultures and yet a considerable amount of cell DNA synthesis. In such a case, the number of cells does not increase, but an autoradiography will show a substantial fraction of cells labeled with [^{3}H]-thymidine. Two explanations are most likely: (1) a number of cells must be dying, which is indeed the case, as we shall see later; or (2) cells increase their DNA amount but do not divide, which also happens in some cases.

DOUBLING TIME

The doubling time is defined as the time required for a cell population to double its number of cells, or for a tissue to double its size. If all cells in the population were dividing, the doubling time would be equal to the length of the cell cycle T_c, and this is what usually happens in cell cultures during the exponential growth phase, when nearly all cells are dividing. But we have seen already that cell populations in vitro and in vivo are a mixture of dividing cells, nondividing cells, and dying cells. The doubling time of a cell population will therefore be the result of cell cycle time, growth fraction (see chapter 3), and rate of cell loss. In tissues, this is further complicated by the heterogeneity of cell types. For instance, in a tumor only about half of the cells are tumor cells while the remainder are fibroblasts, endothelial cells, etc., to which one must also add the intercellular substance. Yet the doubling time of a tumor — the time it takes a tumor to double its mass — is a good measure of a tumor's aggressiveness. Several investigators have determined the doubling times of a variety of human tumors, but the best summary of the data can be found in the book by Steel (1977), who compiled the doubling times of more than 700 primary and metastatic human tumors. A selection from his summary is given in Table 4.2, from which one can see that the doubling times range from 17 days in fast growing Ewing's sarcoma to more than 600 days in primary adenocarcinomas of the colon and rectum. Not included in Steel's tables is the fastest growing tumor of all, Burkitt lymphoma. A child presenting with Burkitt lymphoma is considered by pediatric oncologists to be an emergency requiring immediate treatment, because the tumor is known to double its size over a weekend. Indeed, in a series of Burkitt lymphomas studied by Iversen et al. (1974), the clinical doubling time varied from 38 to 116 hours, with a mean of 66 hours, less than three days. The doubling time of tumors can also explain the latent period of human tumors, i.e., the interval between the carcinogenic insult and the time when a tumor becomes clinically detectable, which is usually when it reaches 1 cm in diameter or ~1 g in weight. Of course in most cases we do not have the slightest idea of when the carcinogenic insult occurred, but some times we do, as for instance with the survivors of Hiroshima and Nagasaki, who developed leukemia with a latent period of 5–8 years. Strong (1977) has reviewed the latent periods of malignancies induced by X-ray therapy in survi-

Table 4.2. Doubling times of primary and metastatic human tumors.

Type of tumor	Mean volume doubling time in days
Primary tumors	
Squamous cell carcinoma of the lung	84
Adenocarcinoma of colon and rectum	632
Carcinoma of the breast	96
Sarcoma of bone	63
Metastases in the lung, from	
Adenocarcinoma of colon and rectum	95
Carcinoma of the breast	73
Ewing's sarcoma	17
Sarcoma of bone	30
Melanoma	53
Lymphoma	27

Adapted from Steel (1977).

vors of childhood cancer. The peak incidence of these radiation-induced malignancies is about 4–6 years, or about 30 doubling times for those tumors with doubling times close to 100 days. However, in patients with the nevoid basal cell carcinoma syndrome, X-ray treatment induces multiple skin tumors within weeks or months. Incidentally, it would be an error to look in a negative way at these iatrogenic malignancies in survivors of childhood cancer. The few secondary tumors that occur after chemotherapy or radiotherapy are a very small price to pay in exchange for the many successes that pediatric oncologists have obtained against a large number of childhood tumors.

Again, these doubling times (and latent periods) are an approximation. Not only do tumors of the same type have different doubling times but the doubling time will vary in the same tumor at different stages of growth. This can be demonstrated especially in tumors of laboratory animals (mice, rats, hamsters), where the growth of a tumor can be followed for the entire life-span of the animal. The size of a tumor especially affects its doubling time, again emphasizing the fact that cell populations, in cultures or in vivo, reach an asymptotic size. Thus, in primary tumors of mice and rats, Steel (1977) found that the doubling times were 5–10 days in tumors of 0.1–

1.0 cm^3 and increased up to 80–90 days in tumors of 10 cm^3 or larger.

Steel (1977) has also pointed out that tumor growth is often irregular, the irregularity varying from periods of rapid growth to stationary volume and even partial regression. These variations in the growth rate of tumors emphasize how careful one has to be in evaluating the effect of anticancer drugs. Without properly matched controls, one can easily mistake for a therapeutic effect what could be simply due to the asymptotic size of a tumor or an irregularity of its growth.

GROWTH OF TUMORS

With this background, it is possible to dissect the growth of tumors into analyzable models that can be reduced to reasonably accurate numbers. As an illustration, we shall study the paper by Bresciani et al. (1974), who investigated a number of squamous cell carcinomas of the skin in man. Selected data from this paper are summarized in Table 4.3.

Let us take, for instance, tumor #2. The cell cycle time is 62 hours. If all cells in the tumor divided (as in the preimplantation embryo), the doubling time of the tumor would also be 62 hours. Instead, it is 312 hours. Clearly, some cells are not dividing. Indeed, the growth fraction is 62 percent. If 62 percent of the cells were to divide with a

Table 4.3. Growth of squamous cell carcinomas of the skin in man. Length of the cell cycle, growth fraction, and doubling time were determined experimentally. Cell birth rate was calculated from the length of the cell cycle and growth fraction. Tumor growth rate was calculated from the doubling time. Cell loss rate is the difference between cell birth rate and tumor growth rate.

Tumor	Length of cell cycle (hr)	Growth fraction	Doubling time (hr)	Cell birth rate[a]	Tumor growth rate[a]	Cell loss rate[a]
1	61	84	288	138	24	114
2	62	62	312	100	22	78
3	52	45	433	86	16	70
4	62	39	600	61	11	50
5	72	45	816	63	8	55
6	88	41	1362	47	4	43

Data summarized from Bresciani et al. (1974).
a. In cells per hr per 10^4 cells.

cell cycle of 62 hours, they would produce 100 cells per hour per 10^4 cell, or a doubling time of 100 hour (for the calculation and the equations used, the reader should consult Table 2 in Bresciani's paper). However, the tumor growth rate (calculated from the doubling time) is 22 cells/hr/10^4 cells. To obtain such a growth rate for tumor #2, one has to assume that 78 cells must die per hour per 10^4 cells.

These figures are obviously only an approximation. As already mentioned, tumor cells constitute only 40–60 percent of all cells in most tumors, the rest being normal cells from the supporting stroma, inflammatory cells, etc. Still, the figures given above are useful in placing tumor growth in the proper perspective, even though the rate of cell loss is probably overestimated. Several conclusions can be drawn from the data of Table 4.3, namely: (1) tumor #3 has the shortest cell cycle but grows slower than tumor #1 and #2; (2) tumor #5 produces more cells per hour but grows more slowly than tumor #4; (3) all parameters—length of the cell cycle, growth fraction, and cell loss—influence the growth rate of a tumor, just as they influence the growth rate of normal tissues.

It is an accepted convention among clinicians that a tumor becomes clinically detectable when it reaches the size of 1 g; obviously this is influenced by the location of the tumor, but it is a reasonable extrapolation. A 1 g tumor contains approximately 5×10^8 cells, of which about 10^8 will be tumor cells. A tumor like tumor #2, producing 100 cells/hr/10^4 cells, 1 g in size, will produce 10^6 new tumor cells every hour. Even if the cell loss rate were correct, the number of tumor cells would be increasing by about 200,000 cells per hour.

Finally, it should be made clear that neither the thymidine labeling index nor the frequency of mitoses are good indicators of the aggressiveness of a tumor, despite many statements to the contrary. The aggressiveness of a tumor is given by its doubling time. Steel (1977), summarizing data from various sources, pointed out that sarcomas, with a median doubling time of 39 days, had a median labeling index of 2 percent, while bronchogenic carcinomas with a median doubling time of 90 days had a median thymidine index of 19 percent. As to mitoses, they last longer in tumor cells than in normal cells (Sisken, Bonner, and Grasch 1982). The increased frequency of mitoses in tumors is partially due to the fact that they last longer than, for instance, in the lining epithelium of the crypts of the small bowel.

Chapter 5

Synchronization of Cells in the Cell Cycle

Synchronization of cells is a very useful device for studying the cell cycle. In exponentially growing populations of cells in culture, and in mitotically active tissues of animals, the cells are distributed asynchronously throughout the cell cycle, the distribution depending on the length of the different phases of the cell cycle. Because metabolic processes vary in different phases of the cell cycle, they can be best studied in synchronized populations of cells. Since this book is not intended to provide details of methodology, the reader who wishes to obtain the technical fine points for cell synchronization must consult the appropriate sources (see, for instance, Terasima and Tolmach 1963; Nias and Fox 1971; Ashihara and Baserga 1979). I would like, however, to discuss briefly the general principles of cell synchronization.

Synchronized populations of cells can be obtained with relative ease in cell cultures. Many researchers have claimed that cells have also been synchronized in living animals. None of these claims has yet to withstand the challenge of rigorous confirmation: either the cells are very poorly synchronized, or synchrony is limited to a small cohort of cells surviving the synchronizing treatment, or both. Occasionally, one obtains surprisingly good results, but they are not reproducible. Although this is purely anecdotal, several years ago we succeeded in labeling, with a single injection of (^{3}H]-thymidine, 95 percent of the basal cells of the esophagus of a mouse, where the labeling index is usually 1–2 percent. This surprising in vivo synchronization was achieved by fasting the animal for 48 hours, followed by re-feeding, and I still have a color slide of the event. Unfortunately, we could never repeat the experiment.

It is regrettable that synchronization in vivo has not yet been obtained. Since the sensitivity of cells to many drugs and physical agents varies in different phases of the cell cycle, synchronization in

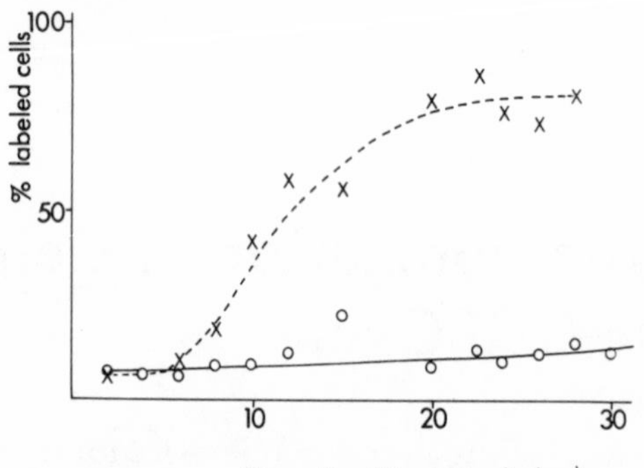

Fig. 5.1 Entry of quiescent (G_0) cells into S phase. Cells made quiescent by serum deprivation were stimulated with 10 percent serum at 0 time. [^{3}H]-thymidine was added at the time of serum stimulation and the percentage of labeled cells at various times after stimulation (cumulative index) was determined by autoradiography. The cells used in this experiment are tsAF8, a temperature-sensitive mutant of the cell cycle that arrests in G_1 at the nonpermissive temperature (chapter 6). Crosses: cells stimulated at the permissive temperature of 34°, Circles: cells stimulated at the nonpermissive temperature of 40°.

vivo would be most useful in maximizing therapeutic effects; but for some reason this problem has never been approached with a serious commitment.

While we are on the topic of unwarranted claims for cell synchronization, I would like to caution the reader that in vitro synchronization is not immune to extravagant claims. For instance, when cells synchronized in G_0 (or in mitosis) are followed as they enter S, one obtains a curve like the one shown in Fig. 5.1, from which it is apparent that, at least in this case, synchronization was modest. The degree of synchronization in these same cells, and by the same method, can be sharply improved by shortening the abscissa and lengthening the ordinate, as shown in Fig. 5.2. This very simple trick has been used quite extensively in the literature; therefore, before adopting a method for cell synchronization, one should look at the figures very carefully.

The degree of synchronization can be monitored by autoradiogra-

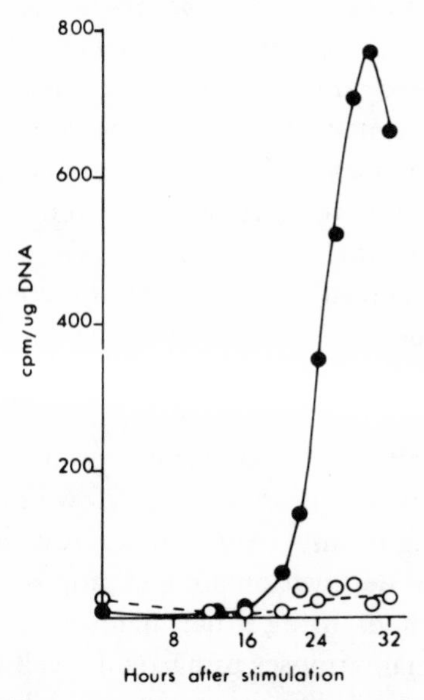

Fig. 5.2 Same experiment as the one describe in Fig. 5.1, except that the incorporation of [^{3}H]-thymidine was a 60 minute pulse and was measured by liquid scintillation counting rather than by autoradiography. Closed circles: cells stimulated at permissive temperature. Open circles: cells stimulated at nonpermissive temperature. (Reprinted, with permission, from Rossini and Baserga 1978.)

phy (or by liquid scintillation counting) after exposure to [^{3}H]-thymidine or by flow microfluorimetry. If one wishes to obtain cells synchronized in the same *phase* of the cell cycle, any one of these monitoring methods is adequate. However, if one wishes to monitor cells synchronized at a *precise point* in one phase of the cell cycle, the autoradiography is the method of choice, and one would then follow the cohort of cells as they enter S (as in Fig. 1.3), or mitosis.

METHODS FOR SYNCHRONIZING CELLS IN CULTURE

Theoretically, cells can be synchronized in any one phase of the cell cycle. However, the most common sites of arrest are M, G_1, and G_0. A G_2 arrest is difficult to achieve, while S phase arrest is usually lethal.

In my opinion, the best method of synchronizing cells is by mitotic detachment, as originally described by Terasima and Tolmach (1963). Since mitotic cells in monolayers are less firmly attached to the glass or plastic surface than interphase cells, they can be collected by shaking the plates gently. With a few discretely spaced shakings, the yield is 5 – 8 percent of the total number of cells, and the percentage of mitotic cells in the collected fraction is 90 percent or more. Colchicine can be used to increase the yield but it is not necessary. The degree of synchrony is excellent, since the arrest is not only phase-specific but almost point-specific. It is convenient for G_1 and S studies, but the degree of synchrony decays as the cells are leaving S. By this method one can easily collect 10^6 cells (>90 percent in mitosis), a number that is sufficient for most biochemical studies. The only shortcoming of the mitotic synchronization method is that it is applicable only to cells that can be grown in monolayers.

For cells growing in suspension, good results have been obtained with the isoleucine deprivation method proposed by Ley and Tobey (1970). Other methods, using either chemical or physical agents, can be useful for special purposes or particular cell lines, but again we must refer the reader to other sources for further information.

MORPHOLOGY OF CELLS DURING THE CELL CYCLE

Synchronized cells are quite useful for studying the morphology of cells during the cell cycle. Apart from striking morphological changes occurring during mitosis (which will not be described here since they are extensively given in textbooks of general biology), morphological changes occuring during the cell cycle are surprisingly modest. Not that descriptions of morphological changes during cell cycle progression are scarce. But most of the reports are anecdotal and are usually valid only for a specific type of cells. Of course, cells grow steadily in size from G_1 to mitosis, but this has already been discussed in chapter 4 and will be dealt again in chapter 9. Apart from the growth in size common to all kinds of cells, perhaps the most spectacular changes are found in lymphocytes stimulated with PHA or other lectins.

Lymphocytes from human peripheral blood are the prototype of G_0 cells, with virtually no cells either in DNA synthesis or mitosis. They are small cells, with a round nucleus, dense chromatin, indistinct nucleolus, scarce cytoplasm—indeed the cytoplasm is reduced to a thin rim around the nucleus. Upon stimulation and before they enter S phase, the cells undergo "blastogenesis" and enlarge several fold. The chromatin decondenses, nucleoli become prominent, and the nuclear membrane shows evaginations (Douglas et al. 1967). The cytoplasm enlarges to a considerable degree, there is a marked proliferation of ribosomes, with an increased number of lysosomes and mitochondria, while the Golgi apparatus becomes highly developed (Inman and Cooper 1963).

While some of these changes can be detected also in other models of stimulated DNA synthesis, they are usually much less dramatic. Not surprisingly, cellular, nuclear, and nucleolar volumes all double from G_1 to G_2. However, it is worthwhile to quote a few numbers from a paper by Lepoint and Goessens (1982). In Ehrlich ascites cells, from G_1 to G_2, cellular volume goes from 880 to 1,725 μm^3, nuclear volume from 325 to 620 μm^3, and nucleolar volume from 49 to 93 μm^3. Ultrastructural cell-cycle specific nuclear and nucleolar changes have also been described in human leukemic lymphoblasts (Parmley, Dow, and Mauer 1977).

Changes in morphology during the cell cycle can be detected by staining monolayer cells with the fluorescent dye quinacrine dihydrochloride (QDH). To quote from the paper by Moser and Meiss (1982): "Early G_1 nuclei are small and usually paired and show a brilliant fluorescence of the chromatin; as the cells progress through G_1, the nuclei appear larger and the fluorescence intensity is reduced; later in G_1 and at the G_1-S boundary, the chromatin shows very low fluorescence except for some bright spots . . . In early S the nuclei are still dark but the intensity increases during middle to late S and continues to increase into G_2, where the chromatin shows a grainy fluorescent appearance." The fluorescence can be quantitated fluorometrically. In 3T3 cells, it decreases from 32 arbitrary units in early G_1 to 11 units in late G_1. Interestingly enough, serum-starved or contact-inhibited 3T3 cells have 28 arbitrary units of nuclear fluorescence, while cells in isoleucine deprivation have only 13.6 units (Moser and Meiss 1982). According to this method, isoleucine-deprived cells are arrested in mid G_1, a conclusion that was also apparent from cell fusion studies using G_1 ts mutants (Tsutsui, Chang, and Baserga 1978).

Another interesting observation is the silver staining of the nucleolus organizer. The nucleolus organizer regions of chromosomes contain rRNA genes, and, when active in producing rRNA, they stain with silver. The number of silver-stained nucleolus organizer regions is at a minimum in G_0 lymphocytes and at a maximum in exponentially growing fibroblasts (Schmiady, Munke, and Sperling 1979).

The centriole cycle is also cell-cycle dependent. The centriole pairs of the cell are replicated once in each cell generation. Separation of the two centrioles occurs during G_1, but replication begins at or near the initiation of DNA synthesis and is completed by G_2 (Robbins, Jentzsch, and Micali 1968). Using indirect immunofluorescence with antitubulin antibodies, Tucker, Pardee and Fujiwara (1979) showed that G_0 3T3 cells have a ciliated centriole. When the cells were stimulated by serum, the centriole's ciliation changed in three phases: first, a transient deciliation; second (by 6–8 hours) a return of the cilium; and third, a subsequent final deciliation of the centriole coincident with the initiation of DNA synthesis.

The morphology of transformed cells can also be discussed in the context of the cell cycle. Again each cell line has its own individual characteristics, but if one wishes to make some generalizations, one can say that transformed cells are usually compact, well separated, and poorly oriented. Untransformed cells are elongated and spindle-shaped and produce colonies of highly oriented cells (Hsie and Puck 1971).

For a while, it was generally believed that transformed cells had an altered microtubular network, but careful studies by Osborn and Weber (1977), using monospecific tubulin antibodies, have shown that the differences are more apparent than real. To quote from their paper: "We conclude that transformed cells contain significant numbers of microtubules, and that in transformed cells, as in normal cells, microtubules are arranged in networks."

The cancer cell can be dramatically different in morphology from its normal counterpart, but for a description of the cancer cell, the reader must be referred to any one of several good textbooks of pathology. But again to point out how elusive are morphological changes, let me say that while some tumors can be diagnosed even by a first-year resident in pathology, others require all the very specialized skills of a competent surgical pathologist.

Part II

Cell Biology

Chapter 6

Temperature-Sensitive Mutants of the Cell Cycle

The usefulness of conditionally lethal mutations for studying structure and function in biological materials has been amply demonstrated by investigations with bacteriophages, bacteria, *Drosophila*, yeasts, and to some extent somatic cells of higher organisms. In theory, therefore, a large number of functions involved in cell cycle regulation should also be amenable to analysis by the isolation and study of conditional lethal mutants, especially temperature-sensitive (ts) mutants. In the past few years, a number of cell-cycle specific ts mutants have been isolated and partially characterized, both in yeast and in mammalian cells. In this chapter they will be described from a kinetic point of view.

In general, the methods used to select temperature-sensitive mutants of the cell cycle are based on the principle that wild-type (wt) cells continue to multiply at the nonpermissive temperature, while the ts mutants are arrested in growth and division. If, after mutagenesis, the cells are shifted up to the nonpermissive temperature and exposed to agents lethal for dividing cells, wt cells should be killed, while ts mutants, being growth-arrested, will survive and grow out again when the cultures are shifted back to the permissive temperature in the absence of lethal drugs (Basilico 1978). This method actually selects for growth, i.e., any mutant that does not grow at the restrictive temperature, while a cell cycle mutant is operationally defined as a mutant that arrests, under restrictive conditions, in a specific phase of the cell cycle. Experience in various laboratories has shown that about 10 percent of ts growth mutants are cell-cycle specific.

The methods for the selection and characterization of cell cycle mutants are actually considerably more elaborate than the brief outline above, and the reader is advised to consult, for details, the papers by Basilico (1977; 1978). From those sources I will summarize a few important points. (1) Among the agents used to kill dividing cells are 5-fluorodeoxyuridine, cytosine arabinoside, high concentrations of high specific activity [^{3}H]-thymidine, 5-bromodeoxyuridine (followed by exposure to UV light), hydroxyurea, and tritiated aminoacids. (2) The selection procedure must be repeated 2 – 3 times, since some wt cells can escape the first exposure to killing agents. (3) A generally long mutation expression time should be allowed before addition of the selective agent. (4) The method does not select for any one specific phase, although G_1 and G_2 mutants are favored, probably because arrest in G_1 and G_2 is more compatible with survival than arrest in S or M. Methods, however, have been developed for selecting ts mutants in a specific phase of the cell cycle (Ohno and Kimura 1984). (5) The method does not select for specific biochemical lesions. Using this methodology, ts mutants of tRNA synthetase were obtained (Thompson, Stanners, and Siminovitch 1975), but unfortunately these mutants are ts for growth and are not cell-cycle specific.

It has often been argued that these mutants should not be called mutants, since there is no rigorous proof that a mutation has occurred in the DNA sequence. This is correct, and indeed we should call these variants, cells with a stably altered phenotype. However, until quite recently very few mutations in higher organisms could qualify for such a rigorous definition of mutants. Recombinant DNA technology and DNA sequencing are changing this, and probably will do the same for cell cycle mutants. For the moment, let us be optimistic and call these cells with a stably altered phenotype ts mutants, and proceed to the definition of a cell-cycle ts mutant.

DEFINING CELL CYCLE MUTANTS

I use the following criteria for the definition of a cell cycle ts mutant: (1) At the permissive temperature, the cells grow normally. (2) At the restrictive temperature, the cells arrest in a specific phase of the cell cycle. For instance, the G_1-ts mutant described in Figs. 5.1 and 5.2, when stimulated with serum from G_0, enters S phase at the permissive temperature (32° – 34°) but not at the nonpermissive temperature (39° – 40°). (3) The point of arrest remains the same even when the cells are shifted to the restrictive temperature in different phases of the cell cycle. Thus, a G_1-ts mutant arrests in G_1 even when the cells are shifted to the nonpermissive temperature during S, G_2, or M. (4) The wild-type cells must be able to grow normally at both permissive and nonpermissive temperatures.

The last point, though trivial, is sometimes forgotten. For instance, my own experience is that BHK cells (Syrian hamster) grow very well at 41°, while at the same temperature CHO cells (Chinese hamster) grow but at a much lower rate than at 37°. Human cells (HeLa) do not grow well above 40°, and 3T3 cells do not grow at all at 39.5°. I do not wish to generalize, since each cell line may have differe. t temperature limits, but in the selection of a wt cell line for the pro^duction of ts mutants one should be aware of these limitations. Clearly, a wt that grows well at 41° gives more flexibility than a wt that dies at 39.5°.

Thus far, I have talked about ts mutants that arrest when shifted to higher temperatures, but there are also cold-sensitive mutants — cells that grow normally at 37° but arrest at a lower temperature (32° – 34°).

CONDITIONAL CELL CYCLE MUTANTS

A partial list of conditional cell cycle mutants of mammalian cells is given in Table 6.1. All phases of the cell cycle are represented,

Table 6.1. Conditional mutants of the mammalian cell cycle.

Mutant (wild type)	Apparent lesion	References
tsJT60 (3Y1 rat cells)	G_0	Ide, Ninomiya, and Ishibashi (1984)
CS4-D3 (CHO) cold-sensitive	G_0/G_1	Crane and Thomas (1976)
tsAF8 (Syrian hamster BHK)	G_1	Burstin, Meiss, and Basilico (1974)
ts13 (Syrian hamster BHK)	G_1	Talavera and Basilico (1977)
Mouse B54	G_1	Liskay (1974)
BF113 (Chinese hamster CCL39)	G_1 (delayed)	Scheffler and Buttin (1973)
H3.5 (Chinese hamster WG1A)	G_1	Landy-Otsuka and Scheffler (1980)
ts11 (Syrian hamster BHK)	G_1	Talavera and Basilico (1977)
tsD123 (rat 3Y1)	G_1	Zaitsu and Kimura (1984)
K12 (Chinese hamster WG1A)	Late G_1	Roscoe, Robinson, and Carbonell (1973)
tsHJ4 (Syrian hamster BHK)	Late G_1	Talavera and Basilico (1977)
ts-2 (Balb/3T3)	DNA synthesis	Slater and Ozer (1976)
dna$^-$ts BN2 (Syrian hamster BHK)	DNA synthesis	Eilen, Hand, and Basilico (1980)
tsA169 (mouse L)	DNA synthesis	Sheinin (1976)
tsT244 (mouse FM3A)	DNA synthesis	Tsai et al. (1979)
ts131b (mouse FM3A)	DNA synthesis	Hyodo and Suzuki (1982)
tsC8 (CHO)	DNA synthesis	McCracken (1982)
ts85 (mouse FM3A)	Late S/G_2	Yasuda et al. (1981)
tsBN75 (Syrian hamster BHK)	G_2 and S	Nishimoto, Takahashi, and Basilico (1980)
ts422E (Syrian hamster BHK)	G_2 and mitosis	Mora, Darzynkiewicz, and Baserga (1980)
ts546 (hamster HM-1)	Mitosis	Wang (1974)
Murine leukemia L5178Y	Mitosis and cytokinesis	Shiomi and Sato (1976)

Table 6.1. (Continued)

Mutant (wild type)	Apparent lesion	References
H6-15 (tsSV3T3)	G_1	Zouzias and Basilico (1979)
ts550C (hamster HM-1)	G_1 and G_2	Chen and Wang (1982)
tsClB59 (mouse FM 3A)	Cytokinesis	Nakano, Sekiguchi, and Yamada (1978)

although G_1 mutants and S phase mutants are the most frequent. One should remember that these mutants are conditionally lethal, and they all die eventually if kept at the nonpermissive temperature. Some of them die quickly; others, like tsAF8, can survive for 3–4 days. The nonpermissive temperature varies with the various mutants. For instance, in short-term experiments (24–30 hours), ts13 can be kept at 39.5°, while tsAF8 should be kept at 40°. Incidentally, the temperature we give in our laboratory is that of the medium in a dish in the incubator, and not the temperature of the air or that of the digital display of the incubator. For longer-term experiments, for instance, selection of revertants or transfectants, the temperature can be dropped 0.5°. The reason we raise the restrictive temperature in short-term experiments is that in a busy laboratory the incubator door is opened often enough to cause a drop in temperature sufficient to allow a ts mutant to partially slip past the block. In long-term experiments, chances are that although a few cells may slip by today or tomorrow, they won't do so the next day. I believe that one should use the lowest temperature compatible with a tight ts block, but it is even more important that the block be tight.

The mutants listed in Table 6.1 are presumably in different complementation groups. (I say presumably because only Basilico and his coworkers have carried out complementation studies on their mutants.) In our laboratory, we have ascertained that tsAF8, ts13, tsHJ4, and K12 belong to different complementation groups. In addition, by fusion of some of these mutants with human cells, it has been shown that the genes correcting three different ts defects reside on 3 different human chromosomes: chromosome 3 for tsAF8, 4 for ts13, and 14 for K12 (Ming, Chang, and Baserga 1976; Ming, Lange, and Kit 1979).

An interesting mutant is H6-15, a 3T3 derived cell line transformed by SV40. The virus (by rescue) could be shown to be wild-type, but the cells are ts for the transformed phenotype; specifically, they have a normal phenotype at 39° and a transformed phenotype

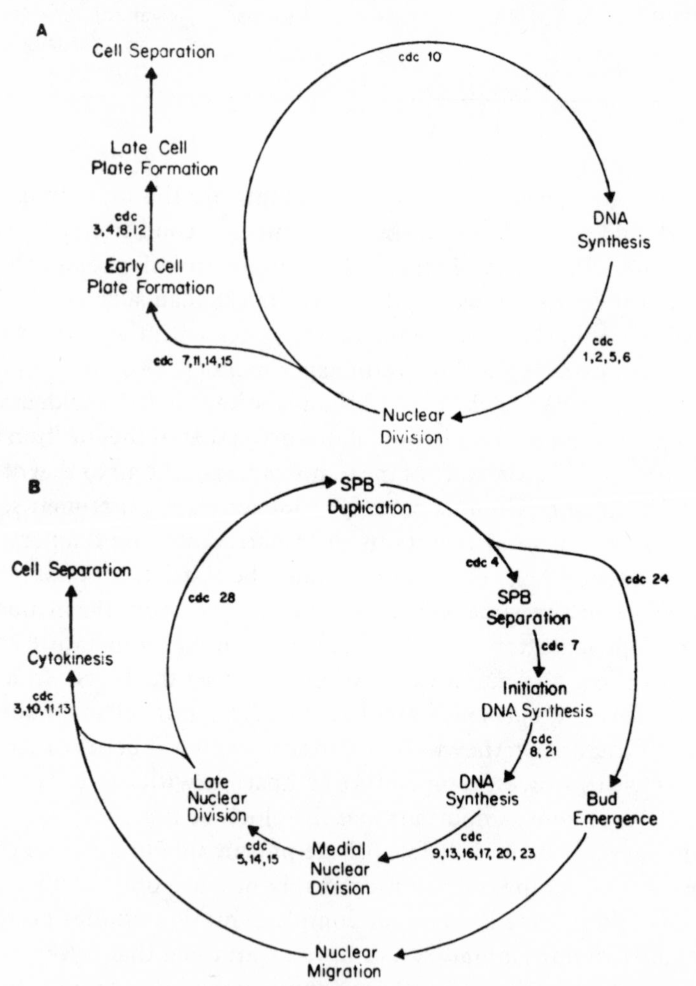

Fig. 6.1 Landmarks in the cell cycles of *Schizosaccharomyces pombe* (A) and *Saccharomyces cerevisiae* (B) derived from mutant phenotypes. cdc designates cell division cycle genes which are located immediately preceding their diagnostic landmark. (Reprinted, with permission, from Hartwell 1978.)

at 32° (Zouzias and Basilico 1979). Since T antigen can also be expressed at 39°, these cells must have a ts cellular function that is necessary for the transforming effect of T antigen. Kinetic studies indicate that at 39° and at a high density, H6-15 cells stop in G_1. We shall return later (chapter 11) to the mutant described by Ide, Ninomiya, and Ishibashi (1984).

We cannot leave the topic of cell cycle mutants without mentioning the beautiful work done by Hartwell with the yeast *Saccharomyces cerevisiae.* Hartwell (1971) took advantage of the morphological changes that occur in the yeast cell division cycle to isolate mutants that are defective in genes that function at specific times in the cycle. Eventually (Hartwell 1976), he isolated a series of mutants, all of which affected the cell division cycle (cdc mutants). By using hydroxyurea and morphological markers, these mutants could be assigned to succeeding steps related in a dependent sequence. Some of these cdc mutants and their position in the cycle of *S. cerevisiae* are given in Fig. 6.1. One step, controlled by a cdc 28 gene product, initiates the cycle, while other gene products control DNA synthesis, nuclear division, and other steps in cell cycle progression. We shall return to these yeast mutants in chapter 13.

THE EXECUTION POINT

Most of this discussion on the execution point is based on the papers by Hartwell (1978) and Pringle (1978), from which I have borrowed freely, although the examples I will use are from mammalian cells and from my laboratory.

While any mutation that blocks cell division could be considered a cell cycle mutation, the term, as mentioned above, is reserved for mutations that lead to a phase-specific block. Classes of mutants may be distinguished from one another, and the role of the defective gene products delimited, by determining the cell cycle stage at which they arrest. The point of arrest can be determined, as a first approximation, in relation to certain cell cycle landmarks which Hartwell (1978) called diagnostic landmarks, and of which several can be easily identified in yeast. For instance, in *Saccharomyces cerevisiae,* spindle pole body duplication, spindle pole body separation, initiation of DNA synthesis, DNA replication, nuclear division, cytokinesis, and bud emergencies can be considered diagnostic landmarks (see Fig. 6.1.) In mammalian cells two easily recognizable diagnostic landmarks are DNA synthesis and mitosis. To quote from Hartwell (1978): "With temperature-sensitive mutants one can determine the point (termed the execution or transition point) in the cell cycle at which the temperature-sensitive event has been completed since, before this point, cells are incapable of dividing upon a shift to restrictive temperature, but after this point they are capable of division." Therefore, the execution point is operationally defined as that time in the cell cycle after which a shift from permissive to restrictive conditions can no longer prevent the mutant cell from successfully completing the current cell cycle (Pringle 1978).

These definitions are quite acceptable if one takes them literally and does not extrapolate them to implications that may or may not be correct. Thus, the execution point does not unambiguously say that a certain protein is no longer needed after a certain stage in the cell cycle. The execution point will depend on whether the mutation is tight or leaky, and it does not distinguish whether the synthesis or the function of the gene product is thermolabile (see Pringle, 1978, for an extensive discussion). For instance, if synthesis of a certain protein is ts and the half-life of that protein is 10 hours, the execution point will be located about 10 hours *before* the true point in the cell cycle where that protein is no longer needed. However, if one takes the definition of the execution point literally, then it is true that it defines the point after which a shift to restrictive temperature no longer prevents progression through the cell cycle. Fig. 6.1 gives a map of the execution point of certain gene products for *Saccharomyces cerevisiae*, but our illustration is taken from mammalian cells.

In the experiments shown in Figs. 6.2 and 6.3, tsAF8 cells (see Table 6.1) were used (Ashihara, Chang, and Baserga 1978). In Fig. 6.2 the cells were made quiescent by serum deprivation; in Fig. 6.3 they were collected by mitotic detachment. Cultures in 10 percent serum were then incubated at either 34° or 40.6° and labeled continuously with [^{3}H]-thymidine. As expected from a G_1-ts mutant, the cells did not enter S at the restrictive temperature, whether stimulated after quiescence or replated after mitosis. At 34° the cells entered S phase normally. In addition, cultures incubated at 34° were shifted up to 40.6° at various times after 0 time and incubated with [^{3}H]-thymidine until the end of the experiment, 32–35 hours in both figures. Two curves of percentage of labeled cells are thus generated: the curve for continuous incubation at 34° and the shift-up curve

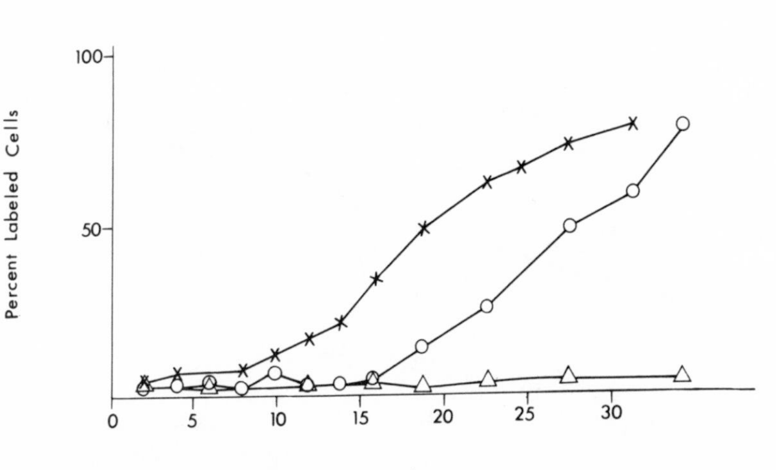

Fig. 6.2. The execution point of tsAF8 cells. Quiescent cultures of tsAF8 cells were serum stimulated at 0 time, when [^{3}H]-thymidine was also added. The percentage of labeled cells at various intervals after stimulation was determined at both 34° (open circles) and 40° (open triangles). In addition, cultures stimulated at 34° were shifted up at the hours (after stimulation) indicated on the abscissa and labeled with [^{3}H]-thymidine until 32 hours after stimulation (crosses). For examples, when cells were shifted to the nonpermissive temperature at 18 hours, 45 percent of the cells had entered S phase by 32 hours. From these 3 curves, one can calculate that the execution point of tsAF8 cells is 8–9 hours before the beginning of S phase (the distance between the curve marked by crosses and the curve marked by open circles; see also text). (Reprinted, with permission, from Ashihara, Chang, and Baserga 1978.)

(we omit the curve at 40.6°, which is purely a control). Because the cells are only parasynchronous, the block is not sharp and the execution point is defined here as the time at which shifting to the nonpermissive temperature no longer prevents 50 percent of the cells from entering S, which is 19 hours after trypsinization and replating of quiescent cells and 3 hours after replating of mitotic cells. However, note that the distance between the two curves is fairly constant, about 8–9 hours. Several growth conditions, besides the ones shown in Figs. 6.2 and 6.3 were tested by Ashihara, Chang, and Baserga (1978). Our conclusions were that (1) the execution point and the time of entry into S varied under different growth conditions; but (2) the time between execution point and time of entry into S was remarkably constant. Thus, regardless of growth conditions, the exe-

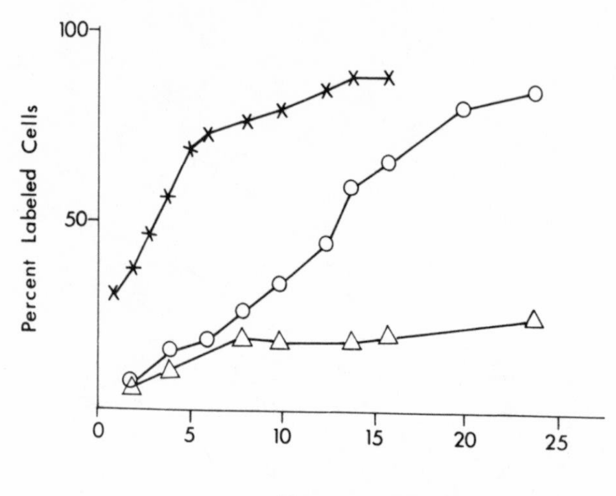

Fig. 6.3. Same experiment and symbols as in Fig. 6.2, except that the cells, collected by mitotic detachment, were replated at 0 time. Notice that the median length of G_1 is 13 hours, against a median of 29 hours for the same cells serum-stimulated from quiescence. Despite the marked time difference of the two conditions, the execution point is essentially the same, 8.8 hours in Fig. 6.2 and 9.7 hours before S in this figure. (Reprinted, with permission, from Ashihara, Chang, and Baserga 1978.)

cution point of the thermolabile function in tsAF8 is located ~8 hours before S. In K12 it is ~2 hours; in ts13 it is ~3.5 hours, always before S.

This is actually a simplification; the real situation is more complicated. Take for instance the mutant tsD123 described by Zaitsu and Kimura (1984). This mutant takes a longer time to enter S phase when it is stimulated from density inhibition than when it is released from its G_1-ts block. That makes sense since, as we have seen, the pre-replicative phase of serum-stimulated cells is almost invariably longer than G_1. However, while tsD123 cells are ts in G_1, they are not ts when stimulated from a density-inhibited state. This would suggest that the G_0 state becomes G_1 at some point beyond the ts block, i.e., in the middle of G_1. Mutant tsD123 would therefore be a true G_1 mutant (just like tsJT60 is a true G_0 mutant), while all the other mutants that stop in G_1 at the restrictive temperature, regardless of whether they come from mitosis or from quiescence, should be called G_0/G_1 mutants.

In some respects the restriction point proposed by Pardee (1974) is an execution point that would fit very well in the model proposed by Zaitsu and Kimura (1984) for tsD123. According to Pardee (1974), cells made quiescent by different methods escape at the same point in G_1 when nutrition is restored. He defined the "restriction point [as] the specific time in the cell cycle at which this critical release event occurs." Pardee's experiments were based on kinetics of cells' re-entry into S from quiescence induced by a variety of conditions.

a. Events occur in a dependent series:

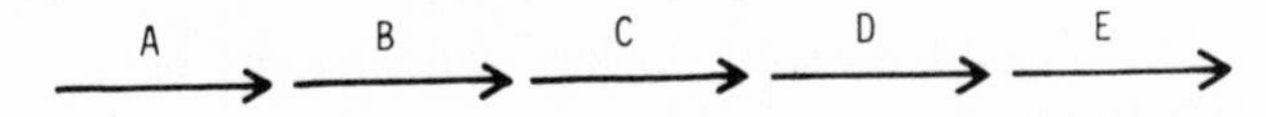

b. Events are independently triggered by some central "clock" (*e.g.*, increasing cell mass):

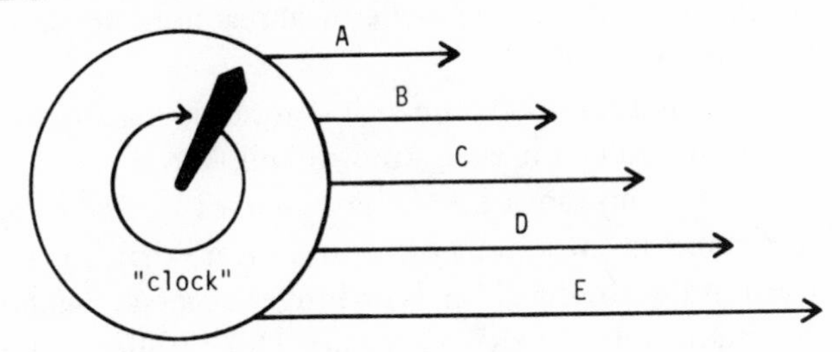

c. Events occur in two parallel pathways, each of which constitutes a dependent series:

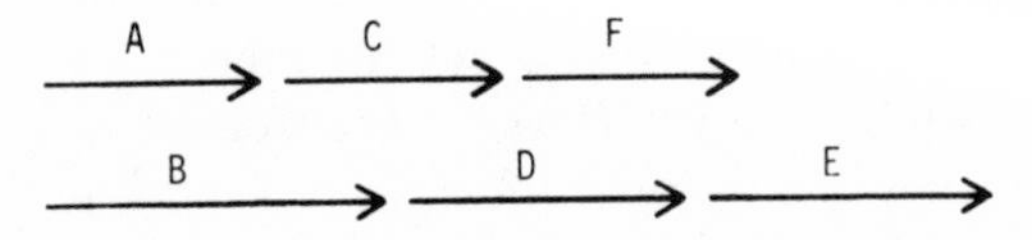

d. Events occur in two parallel pathways (D-E-H and C-F,G-I), each of which is dependent on an earlier event (B), and each of which is prerequisite for a later event (J):

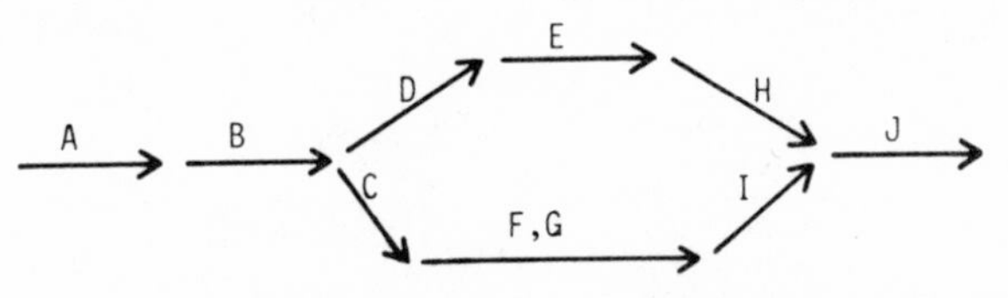

Fig. 6.4 Alternative pathways of cell cycle events. (Reprinted, with permission, from Pringle 1978.)

This, and the experiments illustrated in Figs. 6.2 and 6.3, bring up the point of how orderly cell cycle progression is. Clearly, there is *some* order—the question is how much? Hartwell (1978) felt very strongly about it: "The numerous biochemical and morphological events in each of these processes [cell cycle stages] are undoubtedly rigorously ordered with respect to one another," an opinion which I condivided for a long time. Pringle (1978) was somewhat less confident about the rigidity of this order and offered alternative pathways by which cell cycle events could occur. These alternatives are illustrated in Fig. 6.4. Pringle's opinion was also favored by Levine (1978). But most of these events are in G_1 and we have seen before that, according to Prescott (Liskay and Prescott 1978), G_1 is not essential to the cell cycle but is only a period of growth; there are no G_1-specific events. If true, all that is needed for a cell to enter S phase is for it to reach a certain mass, or a certain amount of some specific proteins; the order in which biochemical reactions are carried out is of less importance.

I still think that there is some order, not universal for all kinds of cells, but some order for each kind of cell. We shall see later that tsAF8 cells are a mutant of RNA polymerase II. The constancy of execution point in these cells seem to indicate that, in tsAF8, transcription of unique copy genes is no longer necessary when the cells reach a certain point in the cell cycle. This implies order. On the other hand, a number of experiments with DNA oncogenic viruses and with alkaline shock (also to be discussed later) raise the question whether the order is rigid or, on the contrary, whether it can be easily circumvented. This aspect of the cell cycle will probably be clarified when the genes controlling cell cycle progression are identified and characterized.

Chapter 7

Informational Content of Cells in Different Phases of the Cell Cycle

Animal cells can spontaneously fuse to form multinucleated cells, as in the formation of myotubes or of osteoclasts during development and foreign body giant cells in certain pathological conditions. Spontaneous cell fusion can also occur in cultures, as first discovered by Barski, Sorieul, and Cornefert (1960), who identified hybrid cells in mixed cultures of two sarcoma cell lines. However, this is a very rare event, whose frequency can be enormously increased by appropriate experimental manipulations. For the following brief introduction to cell fusion, I am largely dependent on the very clear and highly informative book by Ringertz and Savage (1976).

CELL FUSION

There are essentially two ways to induce mass fusion in populations of cells in culture. The first uses UV inactivated Sendai virus and was originally introduced by Okada (1962). The second is based on the use of polyethylene glycol (PEG) and though it was originally discovered by other investigators (see Ringertz and Savage 1976), the present methodologies derive from the technique proposed by Davidson and Gerald (1976) and by Davidson, O'Malley, and Wheeler (1976). In my own experience, PEG-induced cell fusion requires less effort and is easier to control than cell fusion by inactivated Sendai virus. This is not just a personal opinion, since most laboratories engaged in somatic cell genetics have switched to the PEG technique.

The use of cell fusion to study the cell cycle is discussed below. Here, I will limit myself to a few definitions and general aspects of cell fusion. Cells from the same cell line, or two different cell lines, or two different species can be fused. When cells from the same cell line are fused, they produce homokaryons — at first, the two nuclei are distinct (or the various nuclei, if the fusion product is multinucleated) but, if the homokaryon is viable, it undergoes mitosis and nuclear fusion occurs. At any rate "the term homokaryon is reserved for polykaryons formed by fusion of cells among which there is no known disparity in genotype or phenotype" (Ringertz and Savage 1976). The term heterokaryon is used for dinucleate or multinucleate cells formed by the fusion of different cell types. Di-heterokaryons from cells of the same species are usually viable and form hybrid cell lines. Heterokaryons from cells of different species can also form hybrid cell lines, but in such cases, the chromosomes of one of the species are preferentially segregated. In hybrids between rodent cells and human cells, the human chromosomes are usually segregated, but exceptions to this rule have been reported (Croce 1976). Of particular interest for us is the observation that the synthesis of ribosomal RNA of the recessive species (the species whose chromosomes are preferentially segregated) is suppressed. Thus, in the hybrid cells described by Croce et al. (1977) between human and mouse cells, where the mouse chromosomes were segregated, only human ribosomal RNA could be detected, although the mouse chromosomes on which the mouse rRNA genes are located were present. Suppression of gene expression in the chromosomes of the recessive species is not limited to rRNA genes. Lydersen, Kao, and Pettijohn (1980) reported that the expression of several human nonhistone chromosomal proteins is abolished in Chinese hamster – human cell hybrids, where the human chromosomes were being segregated. And in the same human > mouse hybrids mentioned above (Croce 1976), synthesis of mouse histones is suppressed (Ajiro et al. 1978). Of course, many other genes of the recessive species are not turned off, otherwise we would not have somatic cell genetics. Why, in hybrid cells, some genes are turned off and others are not is a question that deals directly with the regulation of gene expression. As a starting point, the reader should consult the thoughtful review by Hohmann (1981).

Fusion can be also obtained between cells and fragments of other cells, especially cytoplasm, the fusion products again being either short-lived (cybridoids) or long-lived (cybrids). Some of these definitions are listed in Table 7.1.

Table 7.1. Definitions of cell fusion products.

Stage	Product	Definition
Immediately after fusion and before nuclear fusion occurs	Homokaryons	Fusion products between cells with the same genotype and phenotype
	Heterokaryons	Fusion products between different cell types; the parent cells can be of the same species or different species
	Dikaryons	Fusion products with two nuclei
	Polykaryons	Fusion products with three or more nuclei
	Cybridoids	Fusion products between a whole cell and the cytoplasm of another cell of different type
Viable fusion products after nuclear fusion	Hybrid cell lines	Permanent cell lines derived from heterokaryons
	Cybrid cell lines	Permanent cell lines derived from cybridoids

In all cell fusion experiments, whether between whole cells or between cells and cytoplasts, it is important to be able to recognize the dual origin of the fusion product. While recognition is easy when fusing chick erythrocytes to mammalian cells, it is more delicate in short-term experiments when fusion takes place between mammalian cells or fragments of cells. A simple method is to label the two different cell lines with beads of different sizes. Beads are easily phagocytized by cells in culture without any disturbance in function. Although some stringent criteria are necessary, this is an easy and useful method (Jonak and Baserga 1979; 1980). More sophisticated methods for determining the identity of hybrid cells have been reported by Hightower and Lucas (1980), who used, among other things, staining pattern with Hoechst 33258, presence of fibronectin, and resistance to azaguanine and diptheria toxin. Of course, if the hybrids are from different species, chromosomal analysis is a very effective method to establish the dual origin of the fusion product. An alternative approach is to select somatic cell hybrids, where the parent cell lines were labeled with two different fluorochromes, by using a two-color flow cell sorter (Junker and Pedersen 1981).

The technique of cell fusion has made possible the rapid development of the whole field of somatic cell genetics. The reader should consult the book by Ringertz and Savage (1976) for the many uses of cell fusion: gene mapping and gene complementation analysis, regulation of gene expression in hybrid cells, analysis of malignancy by cell hybridization, etc. As a technique for the introduction of new information into somatic cells, cell fusion is being partially replaced by recombinant DNA technology and the use of microinjection and DNA transfection. However, it is still a very useful methodology, as illustrated by its application to cell cycle studies.

INFORMATIONAL CONTENT OF S PHASE CELLS

The classic cell fusion experiment in cell cycle studies is the one reported by Rao and Johnson in 1970. These authors fused HeLa cells in different phases of the cell cycle. Their conclusions can be (partially) quoted: (1) "There was obviously a rapid induction of DNA synthesis in the G_1 nuclei of the G_1/S fused cells. The rate of induction was dependent on the ratio of S to G_1 nuclei"; (2) "The G_1 component of the heterophasic G_1/S cell did not inhibit DNA synthesis that was in progress in the S nucleus"; (3) "the G_2 nucleus and cytoplasm had no effect on the normal course of DNA synthesis of an S nucleus"; and (4) in G_2/S cells, there was no re-induction of DNA synthesis in the G_2 nucleus."

The S phase cell, therefore, seems to contain all the necessary information to induce DNA synthesis in G_1 nuclei. Indeed, the S phase cell can even induce DNA synthesis in chick erythrocyte nuclei, although the chick erthrocyte is by all standards a terminally differentiated cell. Two lines of evidence indicate that S phase cells, and only S phase cells, can reactivate chick erythrocyte nuclei: (1) nondividing cells, such as macrophages, cannot reactivate chick erythrocyte nuclei in heterokaryon formation (Ringertz and Savage 1976); and (2) ts mutants of the cell cycle blocked in G_1 by the nonpermissive temperature cannot reactivate, when fused with chick erythrocytes, the chick nuclei (Dubbs and Kit 1976; Tsutsui, Chang, and Baserga 1978). However, if the G_1-ts mutants are allowed to reach S phase at the permissive temperature before fusion to chick erythrocyte, then the chick nuclei are induced to synthesize DNA even when the fusion products are incubated at the nonpermissive temperature (Tsutsui, Chang, and Baserga 1978).

An even more convincing experiment on the informational content of S phase cells is the one reported by Floros and Baserga (1980). When G_0-tsAF8 are fused with other G_0-tsAF8 cells, and the fusion products are incubated at the nonpermissive temperature, the homokaryons are incapable of entering S phase. However, when S phase tsAF8 cells are fused with G_0-tsAF8 cells, both nuclei in the homokaryons enter S phase even at the nonpermissive temperature. Clearly, not only all the information is present in S phase cells, but, at least in the case of tsAF8, a gene product that is needed in G_1 is no longer needed in S phase. This has been confirmed by Mercer and Schlegel (1980), who showed that nuclei of quiescent cells can resume progression toward S phase without the addition of extracellular serum by fusing quiescent cells with cells in late G_1/early S phase. Finally, the critical information of S phase cells is present in the cytoplasm, since cytoplasts of S phase cells can complement other ts mutants (Jonak and Baserga 1980), stimulate G_0 cells to enter S phase in the absence of extracellular serum stimulation (Mercer and Schlegel 1982), and even reactivate chick erythrocyte nuclei (Lipsich, Lucas, and Kates 1978). An indirect confirmation that S phase cells do not need additional information to continue synthesizing DNA comes from the experiments of Roufa (1978) with a Chinese hamster cell line, ts14, temperature sensitive for protein biosynthesis. When ts14 cells in S phase were shifted to the nonpermissive temperature, DNA synthesis continued normally as if all the proteins needed for DNA replication had already been made in G_1. It should be noted, though, that under these conditions the cells were deficient in histones.

All these results lend strong support to the hypothesis of Ringertz and Savage (1976) that the initiation of DNA synthesis in G_1 or G_0 nuclei of heterokaryons depends on the migration into G_1 nuclei of protein factors specific for the S phase. In their absence, as in G_1 arrested ts mutants, other proteins from the donor cell are insufficient, say, for the reactivation of the chick nucleus. Indeed, migration of nucleus- and nucleolus-specific proteins from the donor mammalian cell into the chick nucleus has been repeatedly demonstrated (reviewed by Ringertz and Savage 1976; see also Tsutsui, Chang, and Baserga 1978). However, I would like to suggest a word of caution. The way these data can be interpreted is that the S phase cell (or cytoplasm) contains all the gene products that are required for DNA synthesis per se. It does not rule out other gene products that may now be absent from S phase cells but that were necessary in G_0-G_1 to bring about the induction of the complete DNA synthesizing machinery.

There are exceptions to the rule that S phase cells (or cytoplasts) can induce DNA synthesis in G_0-G_1 cells. When senescent human diploid fibroblasts (see chapter 2) are fused with actively growing human diploid fibroblasts from early passages, the nuclei of the senescent cells are not reactivated. On the contrary, DNA synthesis is inhibited in the nuclei of the young cells (Norwood et al. 1974; Stein et al. 1982), as if the senescent cells had an inhibitor of cell cycle traverse. Interestingly, the inhibitory effect on DNA synthesis by senescent human diploid fibroblasts in heterokaryons applies also to cells transformed by chemical carcinogens or by retroviruses, but not to cells transformed by SV40 or to the irrepressible HeLa cells (Stein et al. 1982). If senescent cells have an inhibitor, it seems that SV40-transformed cells or HeLa cells can neutralize it. Alternatively, these cells may simply by-pass the step blocked by the inhibitor (see Chapter 12). Even young cultures of quiescent human diploid fibroblasts can inhibit DNA synthesis in nuclei of cycling cells when in heterokaryon formation, but only if the quiescent cells have been in G_0 for prolonged periods of time, as reported by Burmer,

Rabinovitch, and Norwood (1984). These authors are careful to point out, however, that a "deep" G_0 is still different from senescence.

A second exception is given by cells treated with butyrate. When G_0 3T3 cells in 3mM butyrate were fused with S phase 3T3 cells, and the fusion products incubated in butyrate-containing medium, the nuclei of the G_0 cells were not induced to enter S phase. However, they were induced to synthesize DNA when fused with SV40-transformed cells (Kawasaki, Diamond, and Baserga 1981).

INFORMATIONAL CONTENT OF G_0-G_1 CELLS

Experiments with G_1-ts mutants offer a good model for an analysis of the informational content of G_0 and G_1 cells. Since these mutants block in G_1 at the nonpermissive temperature, by cell fusion one can truly ask how much cell cycle information do G_0 and G_1 cells contain. The experimental design is simple: G_1-ts mutants are made quiescent by serum deprivation, then they are fused to cells of another type, and the fusion products are incubated at the nonpermissive temperature in 10 percent serum.

First of all, G_0 and G_1 cells cannot reactivate chick erythrocyte nuclei in heterokaryon formation. Three G_1 mutants were studied; K12 (Dubbs and Kit 1976) and tsAF8 and CS4-D3 (Tsutsui, Chang, and Baserga 1978). All of these mutants can induce DNA synthesis in chick erythrocyte nuclei when the fusion products are incubated at the permissive temperature. However, when quiescent ts cells are fused to chick erythrocytes and then incubated at nonpermissive temperature, neither the mammalian nucleus nor the chick nucleus can synthesize DNA. Since CS4-D3 is a cold-sensitive mutant, its nonpermissive temperature is 32°, which rules out the possibility that lack of reactivation of the chick nucleus may be due to the high temperature (39.5°–41°), which is restrictive for tsAF8 and K12. These experiments clearly emphasize the passive nature of chick erythrocyte nuclei reactivation, i.e., chick erythrocyte cannot complement G_1 cells.

On the contrary, when different ts mutants of G_1, made quiescent, are fused, complementation occurs. Thus, when G_0-tsAF8 cells are fused with G_0-ts13 cells (see Table 6.1), the heterodikaryons enter S phase after serum stimulation at both permissive and nonpermissive temperatures (Jonak and Baserga 1979). Complementation also occurs when G_0-tsAF8 cells are fused with cytoplasts from G_0-ts13 cells, indicating that, in the case of tsAF8 cells, the defective gene product is already present in the cytoplasm of G_0 cells. Since the defective gene product in tsAF8 cells is RNA polymerase II (see chapter 9), this is not surprising, as one would expect G_0 cells to contain mRNA coding for RNA polymerase II.

In the case of other mutants, the situation is slightly different. Fusion of G_0 cells does result in complementation, but fusion of G_0 cells with G_0 cytoplasts does not. For instance, cytoplasts from G_0-tsAF8 do not complement G_0-ts13 cells or G_0-tsHJ-4 cells (Jonak and Baserga 1980). Cytoplasts from G_0-tsAF8 cells are also incapable of complementing the defective thymidine kinase in tk^- cells. It seems, therefore, that the information for thymidine kinase and for at least two other gene products needed in G_1 is not present in the cytoplasm of G_0 cells, but requires an active functional nucleus. In other words, "the nucleus is necessary to generate the cytoplasmic appearance of three functions between the time of serum stimulation and the entry of cells into S phase" (Jonak and Baserga 1980). These various results are summarized in Table 7.2.

Similar conclusions can be drawn from an experiment by Smith and Stiles (1981). These authors fused 3T3 cells stimulated with platelet derived growth factor (PDGF) to untreated cells. Under the conditions of their experiment, PDGF does not induce DNA synthesis in 3T3 cells unless plasma is also added (see chapter 10). The fusion products entered S phase by the simple addition of plasma,

Table 7.2. Phenotypic complementation between cell cycle mutants. Phenotypic complementation is defined here as the ability of the fusion products to enter S phase at the nonpermissive temperature.[a]

Fusion products	Phenotypic complementation for DNA synthesis
G_0tsAF8 X chick erythrocyte	–
G_0-tsAF8 X cytoplasts from G_0-ts13	+
G_0-ts13 X cytoplasts from G_0-tsAF8	–
G_0-tsHJ-4 X cytoplasts from G_0-tsAF8	–
tk^- ts13 X cytoplasts from G_0-tsAF8	–

a. All fusion products enter S phase at the permissive temperature. Fusion between cells (rather than cytoplast and G_0 cells) always results in complementation at nonpermissive temperature (except for chick erythrocytes), regardless of whether or not the cells are in G_0. S phase cytoplasts complement all mutants.

which of course was not sufficient to induce DNA synthesis in untreated cells. Cytoplasts derived form PDGF-treated cells were also able to transfer the growth response to untreated cells.

PREMATURE CHROMOSOME CONDENSATION

The best way to describe premature chromosome condensation (PCC) is to quote directly from Ringertz and Savage (1976): "Fusion of a mitotic cell with a cell in interphase results in a precocious attempt of the interphase nucleus to enter mitosis. The chromatin of the interphase nucleus condenses into chromosome-like structures, sometimes with fragmented appearances, and the nuclear membrane disappears." The term PCC was introduced by Johnson and Rao (1970), but, under the name of chromosome pulverization, had been previously described in polykaryons induced by infection with certain viruses, especially measles virus. Chromosome pulverization may in turn be related to thymineless death, i.e., cell death occurring during unbalanced growth. The classic example of thymineless death is given by thymine auxotrophs starved for thymine but not for other nutrients (Cohen and Barner 1954). Using a thymidylate-synthase deficient mutant of mouse cells, Hori et al. (1984) have shown that extensive chromosome breakage follows thymidine starvation. PCC, in this respect, could be considered as a form of acute thymineless death.

From the cell cycle point of view, it is important to know that the appearance of prematurely condensed chromosomes varies with the stage of the cell cycle at which the interphase cells are fused with mitotic cells. This is illustrated schematically in Fig. 7.1, taken from Ringertz and Savage (1976). The conclusive experiments, however, were carried out by Rao and co-workers (see also Rao and Johnson 1974), who fused mitotic HeLa cells with other cells in different phases of the cell cycle. Their results can be summarized as follows: (1) G_1 interphase nuclei produce thin extended single-stranded chromosome filaments; (2) S phase nuclei give rise to irregular, fragmented chromatin masses; (3) G_2 nuclei give thick, double-stranded chromosome filaments; (4) PCC occurs immediately and becomes detectable within 10 minutes after an interphase nucleus is fused with a mitotic cells.

PCC can therefore be used to locate the position of a cell in the cell cycle, and indeed, it can be used to differentiate even cells in early

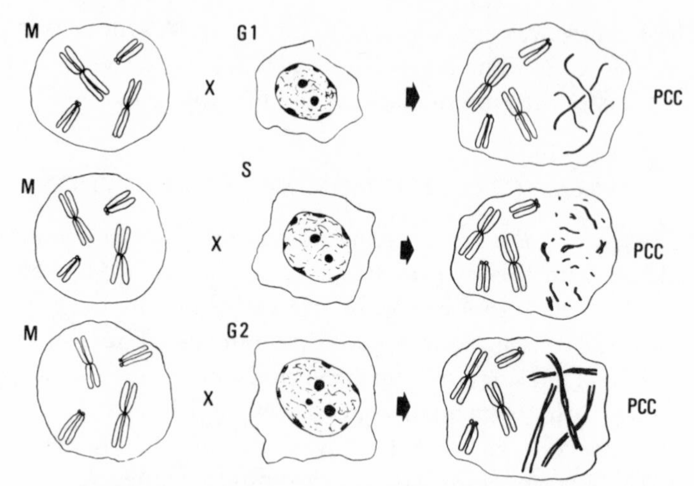

Fig. 7.1 Schematic illustration of the induction of premature chromosome condensation (PCC) in G_1, S, and G_2 cells after fusion with mitotic cells. (Reprinted, with permission, from Ringertz and Savage 1976.)

G_1, mid-G_1, and late G_1 (Rao, Wilson, and Puck 1977). The method is also uniquely suited for the study of noncycling populations or those with a low growth fraction, especially after exposure to cytotoxic agents.

In terms of informational content, it is obvious that techniques like cell fusion, cytoplasts, and PCC are rapidly made obsolete by the dramatic advances in the molecular biology of cell proliferation (see chapter 13). However, I think that these techniques will still be very useful in the future by providing selective methods for localizing, in a time-frame or at a subcellular level, the expression of genes that regulate cell cycle progression. As our knowledge of cell proliferation slowly moves from phenomenology to mechanisms, heterokaryons, cybridoids, and PCC — like autoradiography and cell kinetics — will stay with us but only as part of a battery of manipulations to be used for a deeper understanding of the genetic basis of the cell cycle.

Chapter 8

Perturbation of the Cell Cycle by Drugs

A great variety of drugs have been used, and are still being used, to elucidate the biochemical basis of the cell cycle. A lot of useful information has been generated in this way and is detailed below, but I would like at this point to sound a few words of caution about the use of drugs and to state clearly their limitations. Of course, there is nothing wrong in studying what certain drugs do to cell cycle progression and to determine which phase of the cell cycle is most sensitive to a certain drug. This is basic information that any good pharmacologist would like and ought to have. My concern is with the use of drugs to elucidate basic mechanisms in cell cycle progression.

Most drugs have multiple sites of action, as physicians have regretfully known for centuries. Although they may have one main target, they often have secondary targets which the clinician calls side effects. In terms of our topic, let us take as an illustration actinomycin D, one of the drugs most extensively used in cell cycle studies. Undoubtedly a useful drug, both experimentally and clinically (it is used in the therapy of certain cancers), actinomycin D does not, though, have specificity of action. In the early years we believed that low concentrations of actinomycin D (0.01 – 0.1 μg/ml) specifically inhibited ribosomal RNA synthesis. However, Lindberg and Persson (1972) showed in KB cells that a concentration of 0.04 μg/ml also inhibited the synthesis of mRNA. Then Clark and Greenspan (1979), using 3T3 cytoplasts, showed that serum increased ornithine decarboxylase activity in these enucleated cells, and that actinomycin D added to cytoplasts at the time of stimulation reduced the induction by 90 percent. Thus, actinomycin D also has a cytoplasmic locus of action which complicates the interpretation of results obtained with this drug. However, a negative result is clearly interpretable. No matter how many sites of action actinomycin D has, there is no question that, at high concentrations (2 μg/ml or more), it completely inhibits RNA synthesis. Therefore, if a certain process is *not* inhibited by actinomycin D, one can confidently say that that process does not require previous RNA synthesis. If the process is inhibited, we should limit ourselves to stating that that particular process is actinomycin D-sensitive.

The same conclusions should be applied to most other drugs, but not all. There are some exceptions and, as an illustration, we shall take α-amanitin which, at low concentrations, specifically inhibits RNA polymerase II by binding stoichiometrically to one of its subunits (Cochet-Meilhac et al. 1974). Mutants resistant to α-amanitin can be easily induced; Ingles (1978), on the basis of their behavior and reversion frequency, could conclude that these mutants, capable of growing in 2 μg/ml of α-amanitin, had a point mutation in the α-amanitin binding subunit of RNA polymerase II. Since they grow in the presence of the drug, one can state that, in terms of growth, α-amanitin at low concentrations has one and only one site of action: the α-amanitin binding subunit of RNA polymerase II. All the other effects are secondary.

Other drugs have strict specificity, for instance, ouabain, but the acid test of their specificity requires the development of mutants that are not uptake mutants and whose frequency of reversion suggests a point mutation. This much said, let us see what drugs have taught us in terms of the cell cycle. I will include, among the drugs, monoclonal antibodies. If an antiserum can be used in the therapy of disease, then antibodies enter into the domain of drugs, and monoclonal antibodies, as we shall see, have an exquisite specificity of action.

ACTINOMYCIN D

In 1963 Lieberman, Abrams, and Ove showed that low concentrations of actinomycin D (<0.1 μg/ml) inhibited the entry into S of quiescent rabbit kidney cells stimulated by serum. Actinomycin D did not inhibit DNA synthesis in cells already in S phase, which meant that the drug inhibited some process during the prereplicative phase, that is, the interval between the application of the stimulus to proliferation and the beginning of S. At high concentrations (5 μg/ml) actinomycin D also inhibited the progression of cells from G_2 to mitosis (Kishimoto and Lieberman 1964). In exponentially growing cells, Baserga, Estensen, and Petersen (1965) showed that there is an actinomycin D-sensitive step in G_1 and that, in Ehrlich ascites cells, the last actinomycin D-sensitive step was located about 8 hours before the beginning of S phase. Once a cell had reached that point it became insensitive to actinomycin D (low concentrations) and entered S phase even in its presence. Although literally hundreds of experiments on actinomycin D and the cell cycle have been published since 1965, with few exceptions they have added very little to what could be concluded from the three papers just mentioned. Interestingly enough, both Lieberman, Abrams, and Ove (1963) and Baserga, Estensen, and Petersen (1965) wrongly attributed the effect of actinomycin D to inhibition of ribosomal RNA synthesis. We know better now, but the fact remains that these experiments focused the attention of investigators on RNA metabolism as a regulator of cell proliferation, a concept that can be approached with much more sophisticated techniques today (see chapter 13).

I mentioned above that there are a few exceptions, and one of the exceptions is particularly interesting because of its bearing on what we will have to say later about growth in size and DNA synthesis (see chapters 9 and 11). Laughlin and Strohl (1976) reported that actinomycin D (0.03 μg/ml) had no effect on cellular DNA synthesis induced in BHK cells by infection with adenovirus 2. The same concentration of actinomycin D duly inhibited serum-stimulated DNA synthesis in the same cells. The observation is of interest because, as we shall see later, adenovirus 2 infection can stimulate cell DNA synthesis without stimulation of cellular RNA synthesis (Pochron et al. 1980). Another exception is the elegant experiment of Gordon and Cohn (1971), who studied the induction of DNA synthesis in macrophages fused to melanocytes. Macrophages usually do not proliferate nor do they synthesize DNA, but when fused with rapidly growing melanocytes their nuclei begin DNA synthesis within 2–3 hours after fusion. Using actinomycin D and another inhibitor of RNA synthesis, bromotubercidin, Gordon and Cohn (1971) showed that melanocyte RNA synthesis, but not macrophage RNA synthesis, was necessary for the induction of DNA synthesis in the macrophage nucleus.

One last thing about experiments with actinomycin D. If one takes a population of quiescent cells, stimulates them with serum,

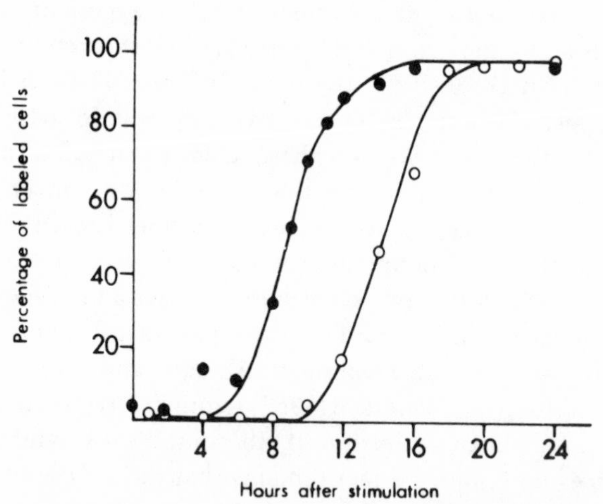

Fig. 8.1 Effect of butyrate on the entry of 3T3 cells into S phase. G_0 cells were stimulated with 10 percent serum and labeled continuously with [^{3}H]-thymidine. The percentage of labeled cells at different intervals post-stimulation is given by the curve with open circles. Other cultures were treated with butyrate (3mM) beginning at the times indicated on the abscissa. All cultures were labeled for 24 hours after serum stimulation. (Reprinted, with permission, from Kawasaki, Diamond, and Baserga 1981.)

and adds actinomycin D at various times after stimulation, the result will look very much like the curve in Fig. 8.1, which simply says that the longer one waits to add actinomycin D, the higher is the fraction of cells capable of reaching S phase. That is almost obvious, considering that the quiescent cells stimulated to re-enter the cycle are not synchronized and the shape of the curve of entry into S is not steep. The longer one waits to add the drug, the more cells have passed the last actinomycin D-sensitive point. It is mentioned here because many drugs give similar inhibition curves and, indeed, the one shown in Fig. 8.1 has been obtained with butyrate. One should not attach any deep significance to these time curves of inhibition.

CYCLOHEXIMIDE

Using synchronized populations of L5 cells, Terasima and Yasukawa (1966) showed that inhibition of protein synthesis in G_1 cells for 2 hours delayed the entry of cells into S by 2 hours. The inhibitor used was puromycin (1 μg/ml) which, however, has a lot of side effects. Many investigators believe that cycloheximide is a better drug for the inhibition of protein synthesis, with fewer side effects and, indeed, cycloheximide has been used extensively to determine the requirements for protein synthesis during the cell cycle. The answer, of course, is simple and obvious: based on the results obtained with cycloheximide, protein synthesis is required for cell cycle progression whether in G_1, S phase, or G_2. As long as we take the effect of cycloheximide with a little grain of salt and remember that cycloheximide may have other effects besides inhibiting protein synthesis, some reports are worth mentioning.

Schneiderman, Dewey, and Highfield (1971), using synchronized Chinese hamster cells, reported that cells exposed to cycloheximide during G_1 were inhibited from entering S. Furthermore, when cycloheximide was removed, they entered S phase with a delay that was greater than the period of exposure to the drug, as if the cells had to start from the beginning of G_1 all over again. Their results suggested a labile protein necessary for the M to S transition, and this was reinforced by a report of Novi and Baserga (1972). These authors used isoproterenol to stimulate cell DNA synthesis in the quiescent salivary glands of mice. The advantage of using isoproterenol was that it is quickly catabolized and disappears from tissues in 30 minutes, so that the proliferating stimulus is very short-lived, although DNA synthesis in the target tissue does not begin until 20 hours after administration of the drug. Novi and Baserga (1972) also used a dose of cycloheximide (33 μg/gm body wt) that inhibited protein synthesis for a period of only 2 hours. With this information they stimulated DNA synthesis in mouse salivary glands with isoproterenol and gave cycloheximide at various times before or after stimulation. When cycloheximide was given in the first 6 hours after isoproterenol, there was complete inhibition of DNA synthesis 27 hours later. If cycloheximide was given 12 hours (or later) after isoproterenol, there was no inhibition whatsoever. To quote from a previous paper from the same laboratory, "Although its [cycloheximide] inhibitory effect on salivary gland protein synthesis has a short duration (2 hr or less), its effect on isoproterenol-stimulated DNA synthesis, when given 1 hr after isoproterenol, is permanent. This would suggest that the template responsible for the synthesis of the 1-hr protein is labile" (Sasaki, Litwack, and Baserga 1969). The suggestion that the templates and the proteins necessary for the stimulation of DNA synthesis have a short half-life, and especially the existence of a labile gene product necessary for cell cycle progression has received further stimulus from the careful experiments of Rossow, Riddle, and Pardee (1979), which will be discussed in chapter 11.

The effect of cycloheximide and other inhibitors of protein synthesis on the progression of cells from G_2 to mitosis has been reviewed by Tobey, Petersen, and Anderson (1971). Save for the reservations mentioned above, it would seem that the last proteins essential for entry into mitosis are synthesized between 30 and 90 minutes before mitosis.

The hazards inherent to the use of drugs are best illustrated by the effect of cycloheximide on on-going DNA synthesis. Cycloheximide and other inhibitors of protein synthesis do rapidly inhibit DNA synthesis in S phase cells (Hodge et al. 1969), from which it was concluded that protein synthesis is necessary for continuing DNA synthesis. But Roufa (1978), using the ts14 mutant of CHL cells, showed that when S phase cells are shifted to the nonpermissive temperature, protein synthesis ceases but DNA synthesis continues normally, which casts serious doubts on the stringency of the relationship between inhibition of protein synthesis and inhibition of DNA synthesis.

OTHER DRUGS

The number of drugs shown to inhibit cell cycle progression is very large. Some of them have even been used as cell-cycle specific drugs in cancer therapy. A review on cell-cycle specific drugs and chemotherapy of cancer can be found in Hill and Price (1982). Table 8.1 simply gives a partial list of drugs that have been reported experimentally to offset cell cycle progression, their point of action in the cell cycle and their presumed target. The qualification "presumed" cannot be overemphasized, and the reader must consult the original articles for the evidence (or lack of evidence) on the specificity of the target. Instead, I will discuss in detail the studies with α-amanitin, which as discussed above, at low concentrations has only one target, the α-amanitin binding subunit of RNA polymerase II.

Wells et al. (1979) reported that α-amanitin (10 μg/ml) added to serum-stimulated AKR-2B mouse cells inhibited hnRNA synthesis, polysomal mRNA accumulation, polyribosome formation, and subsequent DNA synthesis and cell division. Oddly enough, α-amanitin also inhibited, within 4 hours, the accumulation of ribosomal RNA. I say oddly enough because in tsAF8 cells shifted to the nonpermissive temperature, and in which RNA polymerase II virtually disappears, the synthesis and accumulation of rRNA are not inhibited (Ashihara et al. 1978). Baserga et al. (1982) investigated the effect of α-amanitin on cell cycle progression by directly microinjecting the drug into the nuclei of tsAF8 cells. Their results can be summarized as follows: (1) α-amanitin does not directly inhibit DNA synthesis since microinjected cells in S phase continue to synthesize DNA; (2) α-amanitin inhibits the flow of G_1 cells into S; and (3) in tsAF8, the last α-amanitin-sensitive step was located about 6 hours before the beginning of S. This is a direct demonstration that RNA polymerase II-directed gene transcription is needed for the transition of cells from M to S, as it will be discussed later.

Going back to Table 8.1, I wish to point out again some of the problems that can be encountered with drugs. Colchicine and colcemid stimulate DNA synthesis in quiescent cells, according to Otto et al. (1981), but they inhibit it, according to McClain and Edelman (1980). The latter authors pointed out that the colchicine effect varies with the density of culture. D,L-a-difluoromethylornithine has been labeled as a "specific" inhibitor of ornithine decarboxylase and, on that basis, since the drug inhibits the G_1 to S transition, some

Table 8.1. Drugs that affect cell cycle progression.

Agent	Phase affected[a]	Presumed target	Reference
Nitrogen mustard	$G_2 \downarrow$	DNA	Lau and Pardee (1982)
Butyrate	$G1 \neq$	?	D'Anna, Tobey, and Gurley (1980)
W-7	$G_1/S \neq \uparrow$	Calmodulin	Hidaka et al. (1981)
Colcemid	$G_0 \rightarrow S \uparrow$	Microtubules	Otto et al. (1981)
Colchicine	$G_0 \rightarrow S \neq$	Microtubules	McClain and Edelman (1980)
D,L-a-difluoromethyl-ornithine	$G_1 \rightarrow S \neq$	Ornithine decarboxylase	Seidenfeld, Gray, and Marton (1981)
5,6,-dichloro-1-β-D-ribofuranosylbenzimidazole	$G_1 \rightarrow S \neq$	RNA synthesis	Chadwick et al. (1980)
Aminonucleoside	$G_0 \rightarrow G_1 \neq$	RNA synthesis	Cholon, Knopf, and Pine (1979)
α-amanitin	$G_1 \rightarrow S \neq$	RNA synthesis	Baserga et al. (1982)
5-fluorouridine	$G_0 \rightarrow S \neq$	rRNA	Wells et al. (1979)
Histidinol	$G_0 \rightarrow G_1 \neq$	Protein synthesis	Yen, Warrington, and Pardee (1978)
Cordycepin	$S \neq$	DNA (?)	Tomasovic and Dewey (1978)
Hydroxyurea	$S \neq$	Ribonucleotide reductase	Sinclair (1967)
Cytosine arabinoside	$S \neq$	DNA synthesis	Momparler (1972)
Picolinic acid	$G_0 \rightarrow G_1 \neq$	Iron	Fernandez-Pol (1977)

a. ↑ increase; ↓ decrease or delay; ≠ block.

investigators have stated that ornithine decarboxylase induction is an obligatory prerequisite for the entry of cells into S. Although ornithine decarboxylase activity is often increased in proliferating cells (Stoscheck, Florini, and Richman 1980), O'Brien, Lewis, and Diamond (1979), using a tumor promoter, 12-O-tetradecanoylphorbol-13-acetate (TPA), to stimulate DNA synthesis, found that in hamster embryo cells TPA induced ornithine decarboxylase but did not stimulate DNA synthesis, while in human fibroblasts TPA stimulated DNA synthesis but did not induce ornithine decarboxylase.

Even drugs like hydroxyurea and cytosine arabinoside, generally accepted as specific inhibitors of DNA synthesis, turn out to have secondary effects when carefully scrutinized (see the results and discussion in the paper by Mironescu and Ellem 1977). Furthermore inhibition of DNA synthesis by hydroxyurea can lead, upon subsequent release, to gene amplification, as demonstrated in the elegant experiments of Mariani and Schimke (1984).

Drugs can also be used to localize various steps in the cell cycle. As an example we shall take the paper by Wood and Hartwell (1982). These authors used a drug, methylbenzimidazole-20-ylcarbamate (MBC) that inhibits the cell cycle of *Saccharomyces cerevisiae* at a stage subsequent to DNA synthesis but prior to the completion of nuclear division (see fig. 6.1). By using reciprocal shift experiments with a panel of Hartwell's ts mutants, the authors located the MBC-sensitive step after the step defined by cdc 17 but before the steps defined by cdc 14 and cdc 23. Similar experiments with a panel of drugs in mammalian cells led Pardee (1974) to the conclusion that cells arrested in G_1 do so at a common restriction point. Interpretation of such experiments, however, is not easy. By kinetic experiments, Kawasaki, Diamond, and Baserga (1981) found that in tsAF8 cells the butyrate block was located 15 hours before S, while the execution point of the ts block (see chapter 6) is located 8 hours before S. Yet, despite the kinetic differences, the butyrate block and the ts block could not be dissociated, i.e., cells released from the restrictive temperature were still sensitive to butyrate inhibition.

Finally, the reader is again reminded that the cell cycle specificity of drugs (if any) is concentration dependent. An elegant illustration of this point can be found in the paper by Kimler, Schneiderman, and Leeper (1978) in which appropriate concentrations can transform S phase blockers into G_2 blockers.

The skepticism that obviously pervades this section should not make us forget the useful contributions that, in the past, drug studies have made to our understanding of the cell cycle. Perhaps, in this era of purified growth factors, monoclonal antibodies, and cloned genes, the use of drugs to dissect the cell cycle may seem naive, but for a long time it was one of the very few tools at our disposal for such dissection.

MONOCLONAL ANTIBODIES

The decision to discuss monoclonal antibodies under drugs finds its arguable explanation in the fact that antibodies have been used for a long time in the therapy of disease. Indeed, although they have many other uses, monoclonal antibodies have already been used to inhibit human tumor growth (Herlyn and Koprowski 1982). Monoclonal antibodies, because of their high specificity, can be used as specific inhibitors of target molecules that play a role in cell proliferation. In this section we will be mostly concerned with monoclonal antibodies introduced *inside* cells, but they can also be used *outside* cells, as illustrated by the following two reports.

Transferrin is one of the obligatory growth factors for cells grown in vitro (see chapter 10). Trowbridge and Lopez (1982) developed a monoclonal antibody against the receptor for transferrin. Cells grown in the presence of this antibody arrested in the S phase of the cell cycle. Epidermal growth factor (EGF) is one of the best studied growth factors (see below). Schreiber et al. (1981) developed a monoclonal antibody to the EGF receptor. Addition of this antibody to quiescent human fibroblasts stimulated DNA synthesis. According to these authors, their observations "strongly support the notion that the information of the EGF-membrane receptor system resides in the membrane receptor rather than in the hormone molecule and that ligands other than EGF that bind to the same receptor can act as agonists of both the early and delayed responses of EGF."

The effect of monoclonal antibodies on target molecules can also be studied by directly microinjecting them into cells. It has been known for some time that antibodies introduced into viable mammalian cells are not toxic per se (Zavortink, Thacher, and Rechsteiner 1979; Antman and Livingston 1980; Floros et al. 1981) and preserve their specificity of action against the antigen (Antman and Livingston 1980). This is best illustrated in Table 8.2, from which one can conclude that: (1) a monoclonal antibody against the SV40

Table 8.2. Microinjected monoclonal antibodies are nontoxic and have specific action.[a]

Stimulus to DNA synthesis	Antibody microinjected	DNA synthesis
SV40	MC anti-T	Inhibition
SV40	Pre-immune IgG	No inhibition
Serum	MC anti-T	No inhibition
Serum	Pre-immune IgG	No inhibition
Serum	MC Lyt 2.2	No inhibition
pCl-1	Pab 8	Inhibition
pCl-1	Pab 14	No inhibition

From Floros et al. (1981) and Mercer et al. (1983).
a. The antibodies were microinjected at the time of serum stimulation or co-microinjected with SV40 DNA. pCl-1 is a cloned fragment of SV40 that makes a 33K T-antigen lacking the -COOH terminal half. Pab 8 is an antibody (MC) against T that recognizes its NH_2-terminal half, while Pab 14 recognizes its -COOH terminal half. Lyt 2.2 is a control monoclonal antibody.

T-antigen, when microinjected, inhibits SV40-induced DNA synthesis but not serum-stimulated DNA synthesis; (2) pre-immune IgG or Lyt 2.2 (a monoclonal antibody against a cell surface antigen) do not inhibit, when microinjected, either serum-stimulated or SV40-induced cell DNA synthesis; (3) if one induces cell DNA synthesis with a cloned fragment of SV40 that produces a truncated T antigen lacking the -COOH terminal half, such stimulation is inhibited by microinjection of a monoclonal antibody that recognizes the NH_2-terminal half of T antigen, but not of a monoclonal antibody that recognizes only the -COOH terminal half. Thus, microinjected monoclonal antibodies are highly specific and nontoxic. Incidentally, microinjected antibodies have a half-life of ~24 hours (Yamaizumi et al. 1979; Zavortink, Thacher, and Rechsteiner 1979; Mercer et al. 1982). The experiments summarized in Table 8.2 also indicate that cell cycle progression can be inhibited by microinjection of a specific antibody.

In subsequent experiments, Mercer et al. (1982) found that microinjection of a monclonal antibody against the p53 protein—a transformation-related protein thought to have a role in the control of normal cell proliferation (see chapter 11)—inhibited serum-stimulated DNA synthesis in Swiss 3T3 cells. The results are summarized in Table 8.3. Only when the anti p53 monoclonal antibody is microinjected about the time of serum stimulation is cellular DNA

Table 8.3. Effect of a microinjected p53 monoclonal antibody on serum-stimulated DNA synthesis in Swiss 3T3 cells.[a]

Time of microinjection of antibody (hrs after stimulation)	Labeled nuclei (%)[b]		Inhibition (%)
	m.i.	c.	
α p53 (−2)	31.7	60.2	48
α p53 (−0.5)	14.0	70.1	80
α p53 (+2)	46.1	66.2	31
α p53 (+4)	63.5	62.4	none
α p53 (+6)	45.1	46.5	none

From Mercer et al. (1982); Mercer, Avignolo, and Baserga (1984).

a. Swiss 3T3 cells were made quiescent and subsequently stimulated with serum. All cells, microinjected with anti p53 antibodies or not, were labeled with [^{3}H]-thymidine continuously from 0–17 hours. Cells microinjected with a control antibody (Lyt 2.2) gave values undistinguishable from nonmicroinjected controls.

b. m.i. = microinjected; c = control cells surrounding the microinjected area.

synthesis inhibited. Once the cells reach a point in G_1, 4 hours after serum addition, they become refractory to inhibition by the anti p53 antibody. Although there are alternative explanations, these experiments suggested that the p53 protein may be necessary for the exit of cells from G_0.

The problem with microinjected monoclonal antibodies is somewhat the opposite of that with actinomycin D. There, an inhibition meant little but lack of inhibition was informative. With monoclonal antibodies, because of the limited amounts one can microinject, inhibition is of interest but lack of inhibition means nothing. For instance in my laboratory, Ed Mercer microinjected monoclonal antibody against actin into serum-stimulated 3T3 cells. No inhibition was observed but the result could not be interpreted as militating against a possible role of actin in cell cycle progression.

Chapter 9

Growth in Size and Cell Proliferation

We have seen in chapter 4 the evidence that cells grow in size during the cell cycle, that is, all cell components are doubled during the course of the cell cycle. The question I am addressing in this chapter is whether growth in size is necessary for cell division, and what is the relationship between cell size, cell DNA replication, and mitosis. More specifically, I would like to inquire whether an increase in cell size is a prerequisite for cell DNA replication or mitosis or if, instead, the three processes can be dissociated. The question has long-range implications. If the three processes can be dissociated, it would suggest that they are under the control of different signals and one could seriously consider the possibility that diverse growth factors may act, respectively, on cell DNA replication, on cell size, and on mitosis.

We have also seen in chapter 4 that cell size can be measured by either measuring cell mass, or amount of RNA, or amount of proteins. All of them are reasonable and, more important, give the same results. I will add another way of measuring cell size. Ribosomal RNA, in most cells, constitutes ~85 percent of all cellular RNA and is an essential part of ribosomes, where protein synthesis occurs. It is therefore reasonable to assume that rRNA genes may constitute *one* of the targets of environmental signals for growth in size. Therefore, synthesis and amount of rRNA have been considered in the literature and will be considered here as a measure of growth in size.

GROWTH IN SIZE AND CELL DNA REPLICATION

The concept that cell size controls the entry of cells into S phase is an old one. In fact, the first suggestions came from the already cited experiments of Lieberman, Abrams, and Ove (1963) and of Baserga, Estensen, and Petersen (1965). In those experiments, the dose of actinomycin D used presumably inhibited only rRNA synthesis, and the suggestion was made that rRNA synthesis (or accumulation) was a pre-requisite for entry into S phase. Since then a copious literature has confirmed that when cells are growing actively or are stimulated to proliferate, there is an increased synthesis and/or accumulation of rRNA. A long list of references on the subject is tabulated in a review by Baserga (1981). Other evidence is equally suggestive. Killander and Zetterberg (1965) found that cells with a small post-mitotic mass had longer G_1 periods while cells with a larger post-mitotic mass had shorter G_1 periods. More recently, Darzynkiewicz et al. (1979) found a strong correlation between RNA amount and entry into S and stated that "the rate of progression through the cell cycle of individual cells within a population may be correlated with the number of ribosomes per cell." Similar conclusions were reached by Singer and Johnston (1982) working with *Saccharomyces cerevisiae.* They, too, had felt that the number of ribosomes may be crucial, but their experiments indicated rather a role of pre-rRNA metabolism. To quote from them: "Even though our experiments have ruled out ribosome accumulation as the governing factor for cell cycle initiation in yeast, we do feel that it is some aspect of pre-rRNA production that is correlated with the ability to initiate a new cell cycle." Amounts of RNA and proteins also figure to have a prominent role in the model of growth control in *Neurospora crassa* proposed by Alberghina and Sturani (1981). These references have been chosen among many because from them the reader may obtain many more references correlating amounts of RNA and proteins (i.e., cell size) with entry into S. But most convincing of all was the report by Yamashita and Fukui (1980) that in the yeast *Rhodosporidium toruloides*, cells with a temperature sensitive RNA polymerase I are arrested in G_1 at the nonpermissive temperature.

Despite these and many other reports, available evidence clearly shows that in animal cells, growth in size and cell DNA replication can be dissociated. First of all, Fox and Pardee (1970) demonstrated that the heterogeneity in the length of G_1 in mammalian cells is not due to variations in cell size. Neither is it due to nuclear size (Yen and Pardee 1979). There are also several lines of evidence clearly indicating that cell DNA replication and growth in size (as measured by proteins or rRNA yardsticks) can be dissociated. These include the following:

(1) Adenovirus 2 stimulates cell DNA synthesis in infected semipermissive hamster cells (Laughlin and Strohl 1976; Rossini, Weinmann, and Baserga 1979). However, it does not cause an increase in either the synthesis or the accumulation (Fig. 9.1) of rRNA (Pochron et al. 1980), indicating that the adenovirus genome contains sufficient information to stimulate cell DNA synthesis but not to induce growth in size.

(2) The cell cycle duration in NHIK cells is equal to the protein doubling time, but cells can enter S phase even when they have substantially subnormal amounts of proteins (Ronning et al. 1981).

(3) 422E cells are a temperature-sensitive mutant of BHK cells (see Table 6.1) that, at nonpermissive temperature, synthesize 45S pre-rRNA normally but fail to process it, so that new ribosomes are not produced (Toniolo, Meiss, and Basilico 1973). When 422 E cells are serum-stimulated at the restrictive temperature, they enter S phase normally (Mora, Darzynkiewicz, and Baserga 1980) — again ruling out a strict control of ribosome number on the initiation of DNA synthesis.

(4) The SV40 T-antigen coding gene is capable of stimulating cell DNA synthesis in quiescent mammalian cells (see chapter 12). T antigen is also capable of stimulating rRNA synthesis, both in vivo

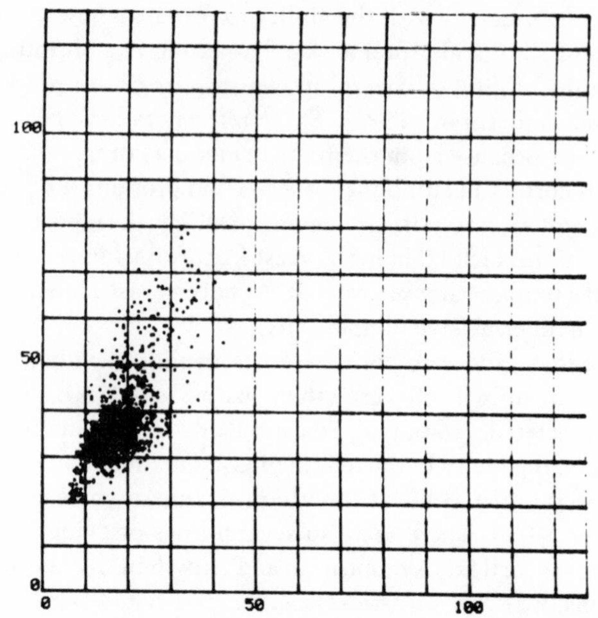

Fig. 9.1 Computer-drawn scattergram of tsAF8 cells infected with adenovirus 2, stained with acridine orange, and analyzed by flow cytophotometry. As in fig 1.6, the abscissa gives the amount of RNA/cell (expressed in arbitrary units of red fluorescence) and the ordinate the amount of DNA/cell (in arbitrary units of green fluorescence). Each dot represents a single cell. Compare with fig 1.6 and notice that cells with >2n amount of DNA (>40 arbitrary units) have very low levels of RNA. Indeed, the amounts of RNA/cell in these S and G_2 cells are lower than in G_0 cells, as one would expect in adenovirus-infected cells, since adenovirus actually inhibits rRNA synthesis. (Reprinted, with permission, from Pochron et al. 1980.)

(Soprano et al. 1983) and in vitro (Whelly, Ide, and Baserga 1978; Learned et al. 1983). However, there are a number of deletion mutants of SV40 that produce truncated T proteins and that are capable of stimulating cell DNA synthesis but not of activating rRNA genes (Galanti et al. 1981; Soprano et al. 1983). Some of these deletion mutants are tabulated in Fig. 9.2. There are at least three mutants in that series that stimulate cell DNA synthesis but fail to activate rRNA genes. Mutants 1001 is one of them, and it makes a T protein 33,000 M.W. instead of the regular 88K T protein of wild-type SV40. Again, cell DNA synthesis and growth in size can be dissociated.

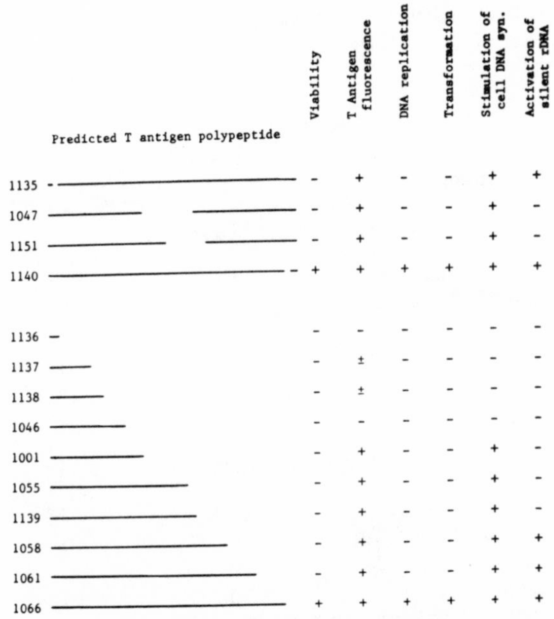

	Predicted T antigen polypeptide	Viability	T Antigen fluorescence	DNA replication	Transformation	Stimulation of cell DNA syn.	Activation of silent rDNA
1135		-	+	-	-	+	+
1047		-	+	-	-	+	-
1151		-	+	-	-	+	-
1140		+	+	+	+	+	+
1136		-	-	-	-	-	-
1137		-	±	-	-	-	-
1138		-	±	-	-	-	-
1046		-	-	-	-	-	-
1001		-	+	-	-	+	-
1055		-	+	-	-	+	-
1139		-	+	-	-	+	-
1058		-	+	-	-	+	+
1061		-	+	-	-	+	+
1066		+	+	+	+	+	+

Fig. 9.2 Biological activities of SV40 deletion mutants. The various deletion mutants of the T antigen coding gene are described in the paper by Pipas, Peden, and Nathans 1983. This figure shows the T polypeptide predicted by the deletion boundaries; in many cases the predicted structure was confirmed experimentally. The wild-type SV40 (not shown here) is of course positive for all the six biological activities listed in this figure. The first column gives the various deletion mutants. Please note that DNA replication in column 5 means *viral* DNA replication, while stimulation of cellular DNA synthesis is in column 7. (Reprinted, with permission, from Soprano et al. 1983.)

(5) When Swiss 3T3 cells are given an alkaline shock (pH 9.5 for 5 minutes), they enter DNA synthesis and divide. However, they do not increase in size, they do not accumulate proteins, and, when they divide, they produce daughter cells that are half the size of the original cells (Zetterberg, Engstrom, and Larsson 1982). Addition of insulin in supraphysiological concentrations (100 μg/ml) caused increase in size and restored balanced growth in the alkaline-treated cells (Table 9.1). Addition of albumin, epidermal growth factor, or transferrin did not cause increase in size (Zetterberg, Engstrom, and Larsson 1982). These findings are particularly important because the dissociation between cell DNA replication and growth in size was obtained without the aid of viruses.

Table 9.1. Effect of different mitogenic stimuli on cellular DNA synthesis, protein content, and dry mass of Swiss 3T3 cells.

Treatment	DNA amount (arbitrary units)	Protein content (arbitrary units)	Dry mass (pg)
BEFORE CELL DIVISION			
None, quiescent cells	20	60	23
10% serum in fresh medium	40	120	47
Alkali shock	40	60	24
Alkali shock + insulin	40	120	—
AFTER CELL DIVISION			
None, quiescent cells	20	60	—
10% serum in fresh medium	20	60	—
Alkali shock	20	30	—
Alkali shock and insulin	20	60	—

Compiled from fig. 4 and 5 and table 4 of the paper by Zetterberg, Engstrom, and Larsson (1982).

(6) When a cloned population of 3T3-4a cells was stimulated by serum, a significant proportion (>30%) of stimulated cells did not show an increase in RNA amount, although they progressed normally toward S phase (Paul et al. 1978).

(7) By microinjection of an antibody against RNA polymerase I it is possible to inhibit nucleolar RNA synthesis and cellular RNA accumulation by 70–80 percent, yet these cells with subnormal amounts of cellular RNA enter S phase normally when stimulated by serum (Mercer et al. 1984).

(8) A functional RNA polymerase II is necessary for the entry of cells into S phase (see below), but it is not necessary for growth in size (Ashihara et al. 1978).

I am omitting from this list the numerous examples in which cells grow in size without entering S phase, which could be considered simply as an incomplete stimulation. The examples given above show that cell DNA synthesis can occur in the absence of growth in size, at least in animal cells. It follows that signals for cell DNA replication can be different from signals for growth in size.

This idea, incidentally, has been around for a number of years. The separation of cell DNA replication from growth in size (nuclear cycle vs. growth cycle) had been brilliantly postulated in 1971 by Mitchison for the yeast cell cycle, but it had been neglected in the field of mammalian cells.

GROWTH IN SIZE AND CELL DIVISION

We have repeatedly noted that growth in size of a cell is an almost self-evident requirement for cell division. Indeed, in the mutant ts422E, mentioned above, in which ribosome accumulation and growth in size are inhibited at the restrictive temperature, entry into S is not affected but cell division is. At the nonpermissive temperature, there is no increase in cell number (Mora, Darzynkiewicz, and Baserga 1980). Similarly, in semipermissive rodent cells, infection by adenovirus causes stimulation of DNA synthesis without growth in size (Pochron et al. 1980), but the stimulated cells do not undergo mitosis (Braithwaite, Murray, and Bellett 1981). And in cells microinjected with the 1001 deletion mutant of SV40 (see above and Fig. 9.2), we observed stimulation of cell DNA synthesis but no mitoses. An illustration of the complex relationship between cell size, cell DNA replication, and mitosis can be found in a report by Malamud et al. (1972). These authors showed that the immunosuppressive agent azathioprine inhibited DNA synthesis in liver cells after partial hepatectomy. It did not, however, inhibit the accumulation in the remaining hepatocytes of RNA and proteins. Furthermore, it did not inhibit some G_2 cells from entering mitosis. Yet, in the experiments of Zetterberg and co-workers discussed above, cell division occurred even in the absence of growth in size.

Despite exceptions, it seems still reasonable to think that, under physiological conditions, growth in size, i.e., doubling of all cell components, is a prerequisite for cell division. Clearly, if it were not so, cells would become smaller at each division and eventually vanish. How can we reconcile, then, the dissociation between cell DNA replication and growth in size or cell division, and yet hold fast to the rule that, under physiological conditions, these three processes must be coordinated? An attractive model that reconciles these apparent contradictions can be found in T lymphocytes stimulated to proliferate. When purified human T lymphocytes, completely free of contamination by macrophages, are exposed to phytohemagglutinin (PHA), they markedly increase in size, the nucleus also increases in size, polysomes reaggregate, and a nucleolus develops. However, PHA-stimulated T lymphocytes do not enter S phase. Addition to PHA-stimulated T lymphocytes of either macrophages or interleukin-2 causes these cells to enter S phase (Maizel et al. 1981). In the absence of previous exposure to PHA, macrophages or interleukin-2 cannot cause growth in size nor stimulate DNA synthesis. An explanation to the puzzle can be found in the fact that, in unstimulated T cells, receptors for T-cell growth factor are masked, unavailable for binding. PHA unmasks these receptors and the T cell becomes now responsive to the T cell growth factor (for a nice review of the subject, see Ruscetti and Gallo 1981).

From this model, one can draw a more generalized model to explain the ordinary coordination among growth in size, cell DNA replication, and cell division. The three processes could be responsive to separate environmental signals. Under ordinary conditions, quiescent cells are responsive only to a signal for growth in size. After the cells grow in size, receptors become available for other environmental signals that are capable of inducing cell DNA replication. It is possible that a third group of environmental signals may be necessary for the induction of mitosis. This would ensure that, under physiological conditions, cells would not synthesize DNA and divide unless they are first stimulated to grow in size.

This model has some interesting corollaries that fit with well-known data. For instance, it would explain the lag between serum stimulation and entry into S, which is always of several hours, as being partially occupied by the time necessary for the unmasking or synthesis of the receptors for the second set of environmental signals. It would explain, of course, all the correlations between cell size and entry into S mentioned above. And it would also explain the phenomenon that has been described as "commitment," about which I will say a few words.

While Todaro, Lazar, and Green (1965) had already observed that it was necessary for confluent monolayers of 3T3 cells to remain in contact with fresh serum for at least three hours to obtain any appreciable stimulation, it was Burk (1970) who demonstrated in BHK cells that the number of cells stimulated to enter DNA synthesis was proportional to the length of serum stimulation. This observation has been confirmed many times, in monolayers and in lymphocytes, and is illustrated in Fig. 9.3, where the cells used were confluent human diploid fibroblasts (Bombik and Baserga 1974). Because

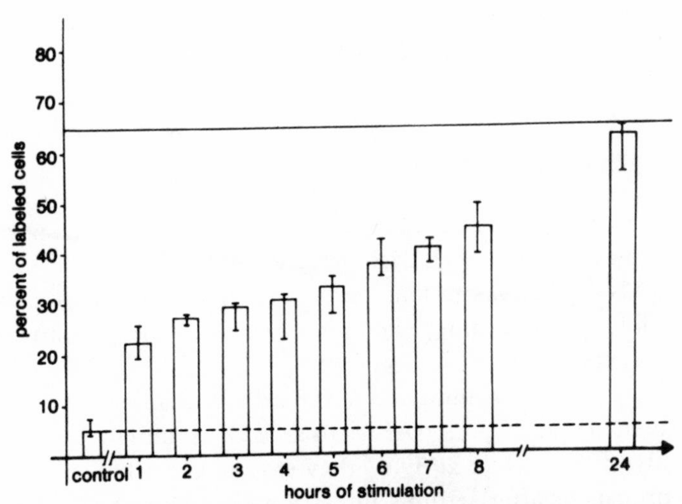

Fig. 9.3 Effect of duration of serum stimulation on the entry of cells into S phase. Quiescent WI-38 human diploid fibroblasts were stimulated with 10 percent serum for the times indicated on the abscissa. Thereafter, the 10 percent serum medium was removed and replaced by medium deficient in growth factors. The cells were continuously labeled with [^{3}H]-thymidine, all cultures were terminated at 24 hours, and the percentage of labeled cells determined by autoradiography. (Reprinted, with permission, from Bombik and Baserga 1974.)

serum is a mixture of growth factors, short pulses of serum may not allow sufficient time for exposure of newly unmasked receptors to the second set of environmental signals. Indeed, with an appropriate timing of exposure to serum, one can even elicit a maximal response in terms of DNA synthesis without stimulating mitoses (van Meeteren, Zoutewelle, and van Wijk 1981), perhaps indicating a need for a third set of environmental signals. Finally, this model would also fit with the already mentioned observation of deeper G_0 states (see chapter 2). We shall return to this topic in chapters 10 and 13, but we may add here that this model is also compatible with the "competence-progression" model proposed by Scher et al. (1979).

If this is true, then the many growth factors that will be discussed in the next chapter should be tested and classified for three separate functions: ability to stimulate growth in size or cell DNA replication or mitosis. The situation does not have to be all or none as in the case of T lymphocytes. For instance, growth in size may make more receptors available for the second set of signals so that the cell becomes responsive to lower concentrations of growth factors. Interestingly enough, many purified growth factors still require minute amounts of serum.

THE ROLE OF RNA POLYMERASE II

There are two main lines of evidence indicating that RNA polymerase II plays a role in cell cycle progression. The first substantial evidence is the demonstration that the transition from G_0 or mitosis to S is inhibited, at the nonpermissive temperature, in a ts mutant of RNA polymerase II.

We have already encountered tsAF8 cells in chapter 6. By all criteria given in that chapter, tsAF8 cells are a G_1-ts mutant. They grow normally at 34°, while at 39.6° they stop in G_1, whether coming from mitosis or after serum stimulation of quiescent populations (see Figs. 5.1 and 5.2). If shifted up after completion of the G_1 period, tsAF8 cells progress through S phase, G_2, and mitosis, and arrest in the next G_1 period. tsAF8 cells are a mutant of RNA polymerase II. Rossini et al. (1980) showed that RNA polymerase II activity decreased sharply in tsAF8 cells at the nonpermissive temperature and that, in fact, the α-amanitin binding subunit of RNA polymerase II actually disappeared with a half-life of about 10 hours (Fig. 9.4). Neither RNA polymerase II activity nor ability to bind α-amanitin were affected in tsAF8 cells at 34°, or in the parent cells, BHK, at either temperature. (We have discussed in chapter 8 how α-amanitin binds specifically and stoichiometrically to a subunit of RNA polymerase II). Subsequently, Ingles and Shales (1982) showed by genetic transfer that the ts defect of tsAF8 cells could be corrected by introducing the gene for normal RNA polymerase II. Finally, Baserga et al. (1982) microinjected into tsAF8 cells a purified preparation of RNA polymerase II, and the cells became capable of entering S even at the restrictive temperature. Although the defect has not been pin-pointed, it can be said that tsAF8 cells have a defect in either the synthesis, the assembly, or the degradation of RNA polymerase II. Therefore, a nonfunctional RNA polymerase II stops cell cycle progression in G_1.

Interestingly enough, tsAF8 cells grow in size normally, at least for 36 hours, at the nonpermissive temperature (Ashihara et al. 1978), and, indeed, RNA polymerase I activity is not affected by the restrictive conditions. It seems therefore that RNA polymerase II is

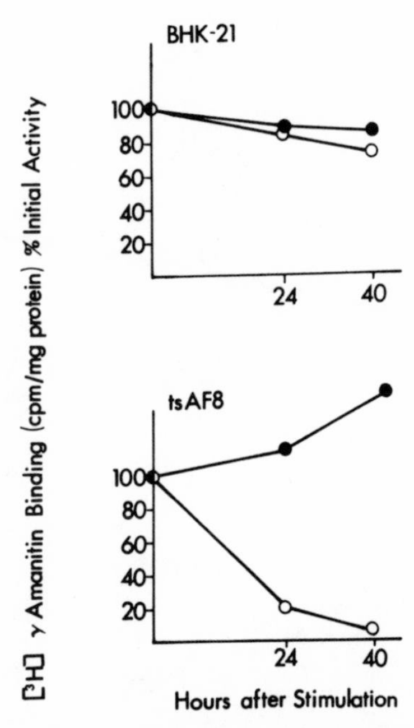

Fig. 9.4 Binding of [^{3}H]γ-amanitin to total proteins of BHK cells (parent cell line) and their mutant, tsAF8 cells. Cells growing in 10 percent serum were used. Closed circles: 32°. Open circles: 40.6°. (Reprinted, with permission, from Rossini et al. 1980.)

necessary for entry into S but not for growth in size, another instance of how growth in size and cell DNA replication can be dissociated.

The need for a functional RNA polymerase II for the entry of cells into S can also be demonstrated by microinjecting α-amanitin directly into exponentially growing cells in culture (see chapter 8). Microinjection of α-amanitin into tsAF8 cells growing at 34° prevents their entry into S (Baserga et al. 1982). We can therefore conclude that RNA polymerase-II-directed gene transcription is necessary for the transition of cells from G_0 or mitosis to S. These experiments served as the basis for the search for genes (cdc genes) regulating cell cycle progression in mammalian cells.

Part III

Molecular Biology

Chapter 10

The Environmental Signals

The extent of cell proliferation in a population of cells is regulated by signals in the environment that are of two general types: spatial restrictions and chemical signals. We have discussed the effects of spatial restrictions in chapters 2 and 3, and I will only add here that spatial restrictions must be operative not only in tissue cultures but also in multicellular organisms. I will deal in this chapter with chemical signals from the environment. That they exist is patently clear to any one who has done the most elementary exercises in cell cultures. Decrease the amount of serum in the medium, and cell proliferation will slow down. Increase the amount of serum, and cells will grow and divide actively. In the whole animal, the presence of regulatory signals for cell proliferation is not so obvious, but it is still evident.

In liver regeneration induced by partial hepatectomy, the removal of $\frac{2}{3}$ of the liver is a modification of the environment, and the salivary gland can be stimulated to proliferate by the injection into the animal of a chemical compound, isoproterenol, that can be considered a growth factor for the salivary gland cells.

The alternative to environmental signals is some kind of programmed cell proliferation, in which the number of cell divisions is limited and automatically determined. We have seen in chapter 2 that the number of cell divisions is indeed limited in human diploid fibroblasts in culture, but there again, for the duration of their life span, the influence of environmental signals is clear. At any rate, this limited number of cell divisions does not seem to play a major role in the intact animal. This topic has been carefully reviewed by Daniel (1977) and I quote directly from him: "Murine cells of several types — hemopoietic, skin and mammary — may display an impressive longevity during the course of serial passage in syngeneic hosts. In several cases the transplants were found to display functional adequacy for time periods well beyond the life span of the laboratory mouse." An elegant experiment that convincingly demonstrates the existence of environmental signals for cell proliferation in the intact animal is the one reported by Lee (1971), who transplanted into rats slowly growing hepatomas. Partial hepatectomy caused increased DNA synthesis and increased growth of the transplanted hepatomas, evidently affected by the same humoral factors that stimulate cell division in the regenerating liver.

The environmental signals that regulate cell proliferation can be more easily studied in the controlled environment of cell cultures; therefore this chapter will largely deal with factors necessary for the growth of cells in vitro. This is not to say that cell reproduction in vivo is regulated by the same factors, but cell cultures are where we must start.

Growth factors can be divided into two large groups: factors that stimulate cell proliferation and factors that inhibit cell proliferation. Both of them are, strictly speaking, growth factors; but following general usage and also for simplicity, we shall refer to stimulatory factors simply as growth factors and use the term inhibitory factors for the second category. The number of known growth factors is much higher than that of inhibitory factors, because it is easier to demonstrate that something stimulates cells proliferation than to show that something inhibits it. In the case of growth factors, the assay gives unequivocal results: the cells either are stimulated to grow or they are not. But in the case of inhibitory factors, we have to exclude in our assay aspecific toxic factors, otherwise HCl would be an inhibitory factor. So, the definition of an inhibitory factor requires not only that it should be pure but also that it should be nontoxic, capable of arresting cells in a specific stage of the cell cycle, and that its action should be reversible.

I am excluding from this chapter the nutrients, inorganic ions, trace elements, etc., that are necessary for the life of a cell. The nutritional requirements of cells in culture were established 30 years ago by Eagle (1955); to say that aminoacids or Mg^{2+} regulate cell proliferation is naive. They are necessary for the well-being of a cell, they are even necessary for cell division, but they do not *regulate* it (see also chapter 11). A meticulous analysis of the growth requirements of cells in culture can be found in Ham (1981). The relationship between low molecular weight nutrients and serum growth factors has been analyzed using a direct kinetic approach by McKeehan and McKeehan (1981). It is useful to quote some of their conclusions from their abstract: "When all nutrient concentrations were optimized and in steady state, serum factors accelerated the rate of multiplication of a normal cell population. The same set of nutrients that supported a maximal rate of multiplication in the presence of serum factors supported the maintenance of nonproliferating cells in the absence of serum factors. Therefore, under this condition, serum factors are required for cell division and play a purely regulatory role in multiplication of the cell population." However, they added: "The quantitative requirement for 18 nutrients of 29 that were examined was significantly higher for cell multiplication in the presence of serum factors than for cell maintenance in the absence of serum factors."

The reader should also note that the amount of growth factor(s) added to a culture is meaningless unless one knows the number of cells in the culture. Since the classic experiment of Holley and Kiernan (1968) on the relationship between amount of serum in the medium and growth curves, it is evident that optimal growth is determined not by the absolute concentration of growth factors in the medium but by the ratio of growth factors to number of cells. Thus 3T3 in sparse cultures can grow even in 1 percent serum, which will not sustain the growth of cells in subconfluent or confluent cultures.

GROWTH FACTORS

Before considering the principal growth factors known, I would like to briefly discuss the growth of cells in serum-free medium. A nice review on the subject can be found in the paper by Sato and Reid (1978), who listed several cell lines capable of growing in completely defined media supplemented by hormones. Since then, many other cell lines have been grown in serum-free media supplemented with growth factors, but the principles remain the same. And the principles are quite simple: (1) all cell lines required insulin and transferrin; and (2) the other requirements varied from one cell line to another. Thus, GH_3 cell required, in addition to insulin and transferrin, parathyroid hormone, tri-iodotyronine, thyrotropin releasing hormone, somatomedin C, and fibroblast growth factor, while HeLa cells required hydrocortisone, epidermal growth factor, and fibroblast growth factor. Several articles on hormones and cell cultures can also be found in the book by Sato and Ross (1979).

Platelet-Derived Growth Factor (PDGF)

My discussion will be largely based on the reviews by Ross (1981) and Stiles (1983). Balk (1971) must be credited with the original observation that serum contains a mitogenic factor that is absent from plasma. Subsequent work from several groups established that the serum factor could be released from platelets. The fact that patients with the gray platelet syndrome have platelets nearly devoid of both alpha granules and PDGF activity indicates that the alpha granules are the source of the serum factor. PDGF, which has been purified, has a molecular weight of 28,000–31,000 daltons and, upon reduction of the disulfide bonds, gives a single band of approximately 13,000 daltons, which, however, is devoid of mitogenic activity. A concentration of 4 ng/ml of the active material is equivalent, in terms of mitogenic capacity, to 1 percent serum. To quote from Ross (1981), PDGF "is a highly cationic, heat stable, disulfide bonded protein of relatively low molecular weight that is present in relatively small quantities in each platelet, but which is sufficiently potent so that with its hormone-like characteristics only small quantities need to be released to stimulate a large number of cells to synthesize DNA."

Cells—for instance, human glial cells—can grow in a defined

medium containing only ng amounts of PDGF (Heldin, Wasteson, and Westermark 1980), provided they are first allowed to attach to the growth surface. However, in most cases PDGF is supplemented with plasma or low amounts of serum. Scher et al. (1979), in a comprehensive review, have listed the factors (in addition to platelet-poor plasma) that complement and potentiate the mitogenic activity of PDGF, and which include somatomedin C. These authors have also formulated an attractive hypothesis that PDGF, incapable per se of stimulating cell DNA synthesis, primed the cells, so to speak, and made them competent to a second growth factor that determined their entry into S phase. Their hypothesis is based on the complete removal of PDGF prior to the addition of the second growth factor. It seems indeed that the binding of PDGF to the cell surface has a half-life of only about 90 minutes (Huang et al. 1982) and that both PDGF and its receptor may be internalized and degraded (Bowen-Pope and Ross 1982). However, the fact that cells can grow in defined medium containing PDGF suggests a more complex model than the one proposed by Scher et al. (1979). One attractive explanation for the competence-progression theory is that cell cycle progression depends on the successive appearance of receptors for different growth factors (Baserga 1984; see also chapter 13).

What is the biological role of PDGF? A number of early cellular responses to PDGF have been described and are summarized in the review by Stiles (1983). These include: tyrosine-specific phosphorylation (see below); inhibition of EGF binding; stimulation of phospholipase A_2 and prostaglandin release; stimulation of polysome formation and of phosphatydylinositol; synthesis and reorganization of actin filaments; increase in low-density lipoprotein receptor content; stimulation of amino acid "A" transport system; stimulation of somatomedin binding (see below); and induction of proteins and new mRNA species. Among these, particularly interesting are the experiments of Pledger et al. (1981) and Scher et al. (1983), who reported that PDGF induces in 3T3 cells the synthesis of specific proteins which are constitutively synthesized in transformed cells. These new proteins appear very quickly (30–40 minutes) after PDGF stimulation; but at least in the case of one of them, the MEP protein, synthesis is induced at lower concentrations of PDGF than are sufficient for DNA synthesis. Scher et al. (1983) concluded that "the synthesis of pII (MEP) may be required, but is not sufficient, for PDGF-modulated DNA synthesis." PDGF also increases the amount of rRNA per cell, but unfortunately in that report (Abelson, Antoniades, and Scher 1979), platelet-poor plasma also caused accumulation of cellular RNA. PDGF also induces phosphorylation of tyrosine in cellular proteins (Cooper et al. 1982). Here, too, though, there is a dissociation between stimulation of DNA synthesis and tyrosine phosphorylation. The new RNA species that are induced by PDGF in quiescent cells will be discussed in Chapter 13. However, mention should be made here, at least briefly, of the startling discovery by Doolittle et al. (1983) and Waterfield et al. (1983) of the homology between PDGF and the product of a retroviral transforming gene, v-*sis*. It raises the intriguing possibility that PDGF may be an autocrine hormone, or, to use the expression of Stiles (1983), "an automitogenic growth factor." Indeed some cell lines derived from animal tumors, including a human glioma, produce a growth factor that has all the characteristics of PDGF (Nister et al. 1984).

Ross (1981) has pioneered the concept that PDGF released by contact of platelets with a damaged vessel wall may be responsible for the stimulation of cell proliferation that occurs in atherosclerosis. In this respect of striking interest is the study carried out by Guzzo et al. (1980) on patients with chronic renal failure on hemodialysis. These patients develop a very rapid, premature, and massive atherosclerosis. Guzzo et al. (1980) measured the plasma levels of a related platelet-derived mitogen, LA-PF4/βTG, in normal individuals (32 ng/ml), in nondialyzed patients with chronic renal failure (135 ng/ml), and in patients on hemodialysis (292 ng/ml). This and the fact that "PDGF-like activity in clotted blood serum coincides with the appearance of the pressurized vascular system on the vertebrate line of evolution" (Stiles 1983) support Ross's concept that PDGF and related proteins may play a major role in the pathogenesis of atherosclerosis.

Epidermal Growth Factor (EGF)

A comprehensive review on EGF, its biochemistry and biological actions, can be found in Carpenter (1981), and that review serves as a basis for the present discussion. EGF was originally purified from mouse submaxillary gland by Cohen (1962) as a low molecular weight protein that, when injected into newborn mice, caused precocious opening of the eyelids and early eruption of the incisors. It soon became apparent, though, that EGF had a mitogenic action in vitro on cultured cells, and most subsequent studies dealt with the in vitro situation.

EGF is a heat-stable, nondialyzable polypeptide containing 53 amino acid residues. Its primary amino acid sequence and location of the intramolecular disulfide bonds (required for biological activity) have been reported by Savage, Hash, and Cohen (1973). The molecular weight is 6,045 daltons. EGF has also been isolated from human urine and from rat submaxillary gland. The amino acid sequence of EGF is 70 percent homologous to that of human urogastrone, a hormone that causes a decrease in gastric acid secretion. EGF is also the major growth promoting factor in human milk (Carpenter 1980), and both human and bovine milk can support the growth of cell lines in culture (Sereni and Baserga 1981).

The addition of EGF (5 – 10 ng/ml) to quiescent cells leads to activation of the cells and their entry into S phase. This stimulation is accompanied by the usual spectrum of biological phenomena that accompany the mitogenic response (increase in sugar transport, cation fluxes, ornithine decarboxylase activity, protein synthesis, etc., discussed in detail in chapter 11). Stimulation of DNA synthesis in cultured cells by EGF requires the presence of low levels of serum (0.5 – 2 percent) and, under these conditions, optimal amounts of EGF can cause an increase in the number of cycling cells from 0.9 percent to 21 percent.

EGF interacts with specific, saturable receptors that have been demonstrated in a wide variety of cultured cells. According to Carpenter (1981): "Two points relating EGF binding to maximal stimulation of DNA synthesis are reasonably clear: (1) persistent interactions between EGF and cell surface receptors must occur for many hours and (2) occupancy of approximately 25% of the available binding sites is necessary." Less clear is the role of internalization and degradation of EGF in stimulating DNA synthesis. The most recent evidence seems to indicate that internalization is not necessary for stimulation of DNA synthesis (Carpenter 1981). Indeed, using cultured bovine granulosa cells, Savion, Vlodavsky, and Gospodarowicz (1980) obtained a full mitogenic response in the presence of 90 – 95 percent inhibition of EGF degradation.

I have mentioned above that EGF induces the usual spectrum of biological responses that accompany stimulation of cell proliferation. Most of these changes (activation of $Na^+ - K^+$ ATPase, changes in microfilaments and cell morphology, activation of ornithine decarboxylase) can also be induced by a nonmitogenic analogue of EGF, a cyanogen bromide-cleaved EGF (Yarden, Schreiber, and Schlessinger 1982). This applied also to tyrosine phosphorylation. As is the case of PDGF, EGF induces tyrosine phosphorylation, probably through a membrane-localized receptor-kinase (Carpenter, Stoscheck, and Soderquist 1982), but again there is a dissociation between tyrosine phosphorylation and mitogenesis, as indicated in the discussion of the paper by Cooper et al. (1982).

Proteases

Burger (1970) was the first to report that proteases could stimulate cell proliferation. Addition of trypsin or pronase to quiescent, contact-inhibited 3T3 cells caused one round of cell division. Since then, numerous reports have appeared indicating that a number of proteases can induce DNA synthesis and cell division in a variety of cell lines. According to Cunningham (1981), whose review constitutes the basis of this discussion, other proteolytic enzymes that have been reported to induce DNA synthesis include thrombin, plasmin, elastase, and the γ-subunit of nerve growth factor. Cell types vary from chick embryo fibroblasts (the most sensitive of all to proteases) to 3T3, human fibroblasts, and even B lymphocytes, although T lymphocytes seem to be refractory.

It should be pointed out that the stimulation of cell division (or DNA synthesis) by proteases is far from being spectacular, especially in mammalian cells. In fact, as discussed by Cunningham (1981), the initiation of 3T3 cell division by trypsin is tricky and quite difficult to repeat. Thrombin, in optimal concentrations (250 nM), induces DNA synthesis in only 10–20 percent of nondividing cell populations (Glenn et al. 1980), less than can be obtained, for instance, with PDGF or EGF. Proteases, though, are interesting because they offer a unique lead to the mechanism by which they stimulate cell proliferation. Some of the most important aspects have been summarized by Cunningham (1981): (1) the enzymatic activity of mitogenic proteases is required for stimulation of cell division; (2) immobilized proteases, for instance, thrombin bound to Sepharose beads, can still stimulate cell division, indicating that they act at the surface of the cell and that internalization is not necessary; (3) receptors have been identified on responsive cells, at least for thrombin; these receptors are specific, since thrombin binding is not competed by EGF or insulin, although it is competed by plasminogen activator; (4) a number of cell surface proteins are cleaved by mitogenic proteases and this proteolysis of the cell surface seems to be required for stimulation of cell division. More difficult is to find a role for protease-induced cell proliferation in vivo, although Chen and Buchanan (1975), who were the first to report thrombin-induced DNA synthesis, suggested that thrombin may play a role in tissue repair following injury.

Growth Hormone, Somatomedins and Insulin-like Growth Factors

There is no question that growth hormone (GH) is one of the principal regulators of balanced postnatal growth in vivo. In utero, growth is probably regulated by an embryonic somatomedin, different from adult somatomedins (Sara et al. 1981). The role of GH in postnatal growth has been known for a long time and can be summarized in the clinical observations that pituitary dwarfs have low serum levels of growth hormone, while high levels of GH are present in patients with acromegaly and gigantism. Indeed, administration of GH to pituitary dwarfs is followed by dramatic increases in cell number and cell size (see the review by Clemmons and Van Wyk 1981). Experimentally, Palmiter et al. (1982) obtained even more startling results. They first constructed a hybrid gene, consisting of the promoter of the mouse metallothionein I gene fused to the (promoterless) structural gene of rat growth hormone. The hybrid gene was microinjected into the pronuclei of fertilized mouse eggs. Seven mice were obtained that carried the fusion gene. Since the metallothionein I promoter is a strong promoter that is constitutively expressed, some of these mice had high serum levels of GH, and six out of seven grew significantly more than their siblings not carrying the fused gene. A beautiful review on the behavior of the metallothionein – human growth hormone hybrid gene in transgenic mice (i.e., mice originating from fertilized mouse eggs microinjected with the hybrid gene) can be found in the paper by Palmiter et al. (1983). One very interesting feature of these transgenic mice is that, although the hybrid gene is present in all tissues, it is well expressed in some organs, like liver and testis, but not in other organs, for instance, kidneys. It seems to provide an attractive example of organ-specific regulation of gene expression. This review also shows that there is not a perfect correlation between levels of GH and overall growth, indicating that other factors must also play a major role in postnatal growth.

Closely related to GH are prolactin and chorionic somatomammotropin (placental lactogen). These three hormones have a high degree of amino acid homology and overlapping biological activities (Baxter et al. 1979). The genes for GH and prolactin are expressed by the pituitary, while chorionic somatomammotropin, as its second name indicates, is expressed by the placenta. Prolactin is a required growth factor for certain cell lines (Sato and Reid 1978) and, as we shall see in chapter 13, one of the cell division cycle genes has extensive homology with prolactin. Although GH is active in vivo, when Salmon and Daughaday (1957) tried it in vitro it turned out to be inactive, the growth-promoting activity being restored by plasma from normal or GH-treated hypophysectomized rats. This finding led to two consequences: (1) the discovery of a family of closely related peptide growth factors that mediate the action of GH in vivo and were called somatomedins; and (2) ironically, a certain degree of neglect of GH itself, since most investigators prefer to work with cells in culture, where GH is ineffective. This neglect has an added twist, since recent experiments by Madsen et al. (1983) indicate that GH can actually directly stimulate DNA synthesis in cultures of chondrocytes. Despite these recent results, somatomedins have some interesting biological activities that are worth mentioning. The following discussion is largely based on the review by Clemmons and Van Wyk (1981).

Somatomedins constitute a heterogenous group of peptide growth factors. Three major characteristics distinguish somatomedins from other growth factors: (1) their serum concentrations are growth-hormone dependent; (2) they are mitogenic for a variety of cell types; and (3) they produce insulin-like actions. The somatomedins in the human plasma which have been best characterized include: (1) somatomedin C (SM-C) and insulin-like growth factor I (IGF-1) that are now considered to be identical, basic peptides (Van Wyk, Russell, and Li 1984); (2) insulin-like growth factor II (IGF-II), clearly a different gene product, since only 70 percent of the residues are identical with those of SM-C/IGF1; (3) somatomedin A (SM-A). In addition, multiplication stimulating activity (MSA) produced by rat hepatocytes cultures belongs to the family of somatomedin. The rest of this discussion will largely deal with SM-C.

According to Van Wyk et al. (1981), in patients with acromegaly and hypopituitarism "clinical assessments of disease activity correlate far better with blood levels of SM-C than they have with growth hormone concentrations." However, our main interest in this section is in the biological activity of somatomedins in vitro and, specifically, their ability to stimulate cell DNA synthesis and cell growth. SM-C has been demonstrated to stimulate [^{3}H]-thymidine incorporation and mitosis in chick embryo fibroblast and other cell types, including human fibroblasts. The optimal concentration for mitogenic action is in the range of 5–10 ng/ml, a concentration which is below those that occur in human serum. The other somatomedins mentioned above also have mitogenic action, especially on chick embryo fibroblasts. Even more interesting, though, is the interaction of SM-C with other growth factors necessary for cell cycle progression. SM-C is apparently the active principle in the platelet-poor plasma (PPP) preparations that act synergistically with PDGF. Clemmons and Van Wyk (1981) offered an attractive explanation for the ability of PDGF to make cells responsive to SM-C. To quote from them: "Exposure of quiescent Balb/c 3T3 fibroblasts to PDGF followed by somatomedin C-deficient PPP resulted in a two fold increase in the specific binding of ^{125}I-somatomedin C to these cells. This increase was accounted for by an increase in receptor number." We have seen in the previous chapter the T lymphocytes model, in which PHA, nonmitogenic per se, makes the cells capable of binding T cell growth factor, which, in turn, causes the T lymphocytes to enter S phase. A similar mechanism could be operative in Balb/c 3T3, i.e., PDGF makes them responsive to SM-C by increasing the number of receptors, and thus allow SM-C to stimulate cell DNA synthesis. This may also explain why human fibroblasts respond more readily to PDGF than Balb/c 3T3 cells in terms of full mitogenic stimulation. Human fibroblasts are stimulated by GH or by PDGF (but not by EGF) to produce somatomedin-like peptides, which they release in the medium, where they act as autocrine hormones (Van Wyk et al. 1981). The production of somatomedins by human fibroblasts depends on numerous variables, especially cell density and other factors (Clemmons and Shaw 1983).

Insulin

A review and a good analysis of the role of insulin as a growth factor can be found in Gospodarowicz and Moran (1976). There is, of course, no question that insulin should be considered a growth factor: as mentioned above, in serum-free media it is an obligatory component for all cell lines (Sato and Reid 1978). The question is whether insulin is mitogenic per se. It is a fairly potent mitogen for chick embryo fibroblasts (Vaheri, Ruoslahti, and Hovi 1974) and, under certain conditions, for rat hepatoma cells (Massagué, Blinderman, and Czech 1982). In the latter system, insulin induces cell DNA synthesis at physiological concentrations (10^{-9} nM) and is 100 times more potent than IGF-II. However, for most mammalian cells, insulin is, at best, a poor mitogen, requiring pharmacological concentrations for rather modest effects. The most attractive explanation for the role of insulin as a growth factor is the one proposed by Gospodarowicz and Moran (1976): "Although insulin is probably a poor mitogen . . . its role in cell division is far from negligible, since in a number of cell systems, insulin potentiates the effect of other growth factors although it has little effect by itself. This probably results because insulin is an anabolic agent required to keep the cells healthy and fully responsive to mitogenic stimuli."

A striking example can be found in the experiments of Zetterberg, Engstrom, and Larsson (1982), already discussed in chapter 9. In those experiments, a 5-minute treatment of Swiss 3T3 cells with alkaline medium (pH 9.5) caused the cells to enter S phase. However, the cells did not grow in size and, upon mitosis, they generated cells half the size of the original cells. Addition of insulin (100 μg/ml) restored balanced growth, i.e., the alkali-shocked cells grew in size and, after division, produced normal-sized cells (see Table 9.1).

One can easily visualize insulin as a hormone that causes an increase in the size of a cell and makes receptors for other growth factors available for their mitogenic action (see above for SM-C and also the role of PHA in making T lymphocytes responsive to T cell growth factor). Insulin is known to stimulate both RNA and protein synthesis (Vaheri, Ruoslahti, and Hovi 1974) and to stimulate cellular phosphorylation in general, the phosphorylation of ribosomal protein S6 being the most spectacular example (Smith, Rubin, and Rosen 1980). A possible mechanism for the potentiating action of insulin on the mitogenicity of other growth factors can be found in the experiments of Renkawitz et al. (1982), who showed that the addition of insulin (5 μg/ml) markedly increased the expression of genes microinjected into mammalian cells.

Finally, it should be pointed out that the mitogenic action of insulin may vary, depending on the species from which the hormone was isolated. Thus, King, Kahn, and Heldin (1983) found that insulin from hystricomorphs (such as guinea pig and porcupine) is a more potent inducer of DNA synthesis in human fibroblasts than other mammalian insulins or IGFs.

Transforming Growth Factors

In 1978 DeLarco and Todaro partially purified a growth factor from serum-free media conditioned by 3T3 cells transformed by murine sarcoma virus. They called it sarcoma growth factor (SGF). The most interesting property of SGF was that it made rat fibroblasts anchorage-independent, i.e., capable of growing in soft agar. DeLarco and Todaro (1978) proposed that the transformation by murine sarcoma virus involved the production of SGF, and that the producing cells were able to respond to their own growth factors.

Since then, transforming factors (TFs) capable of rendering normal cells anchorage-independent have been identified in the media conditioned by a variety of transformed cells. Thus, Ozanne, Wheeler, and Kaplan (1982) identified TFs produced by cells transformed by Kirsten murine sarcoma virus, by SV40, and by polyoma, while Moses and Robinson (1982) found that TF activity was produced by cells transformed by 3-methylcholanthrene. With Sporn et al. (1983) we can say that "transforming growth factors are a heterogenous set of low molecular weight polypeptides defined by their ability to induce the transformed phenotype—particularly anchorage-independent growth in soft agar—in untransformed indicator cells that ordinarily do not grow in soft agar." TFs have now been isolated also from normal tissues, such as bovine salivary gland and kidney (Sporn et al. 1983). These authors have also proposed that TFs may play a role in wound healing.

There are two general classes of TFs, based on their relationship with EGF. Type αTFs bind to EGF receptors, are isolated from the conditioned medium of transformed cell lines, and cause anchorage-independence by themselves. They are structurally as well as functionally related to EGF and human urogastrone (Marquardt et al. 1984). In contrast, type βTFs do not bind to EGF receptors, but they require EGF or TF-α to induce the transformed phenotype (Assoian et al. 1983). Apart from their possible role in wound healing, TFβ is also interesting because it controls the level of EGF receptors (Assoian et al. 1984), a mechanism which, as we have seen, may play a major role in cell cycle progression.

Hemopoietic Colony Stimulating Factors

We have already seen in chapter 3 how the concept of stem cells originated from the experiments of Till and McCulloch (1961), who obtained hemopoietic colonies in the spleen of irradiated mice injected intravenously with syngeneic bone marrow cells. The subsequent development of methods for growing hemopoietic colonies in vitro has allowed the identification and characterization of specific regulators for the proliferation and differentiation of hemopoietic cells. In this brief and greatly simplified discussion of those regulators, I am following quite closely the review by Metcalf (1981).

The term "colony stimulating factor" (CSF) denotes the active factor(s) in the environment that promote the growth of hemopoietic cells, and, operationally, it includes not only purified CSF but also crude or incompletely characterized stimulating material. An important characteristic of hemopoietic CSFs is that they not only promote growth, but they do so along well-defined lines of differentiation. Metcalf (1981) distinguishes the following CSFs: (a) *Granulocyte-macrophage colony stimulating factor* (GM-CSF), which is produced by a variety of cell types (macrophage, fibroblast, mitogen-activated T lymphocytes, endothelial cells) and organs. Different GM-CSFs have been extracted from human and mouse tissues, their molecular weights ranging from 23,000 to 70,000. The best characterized GM-CSFs seem to be glycoproteins; for instance, the one from mouse lung is a neuraminic-acid containing glycoprotein, 23,000 in molecular weight, where carbohydrates comprise 10–20 percent of the molecule. As the name says, GM-CSF promotes the proliferation of all cells in the granulocyte-macrophage pathway from the earliest committed progenitor cell to the last cells of the series capable of proliferation, myelocytes and promonocytes. This growth stimulation eventually results in the progressive differentiation of most of the colony cells to mature, postmitotic polymorphonuclear leukocytes and macrophages. (b) *Eosinophil colony stimulating factor* (EO-CSF), which promotes the growth of loose, dispersed colonies of eosinophils. In mouse, such an activity is generated by the interaction between T-lymphocytes and macrophages, but in human, it is produced by unfractionated peripheral blood cells and by the placenta. (c) *Megakaryocyte colony stimulating factor*

(MEG-CSF), a glycoprotein of molecular weight 24,000 probably produced by activated lymphoid cells. (d) *Erythropoietin and erythroid colony stimulating factor.* Purified erythropoietin is a neuraminic-acid-containing glycoprotein of molecular weight 39,000. It has distinct actions on target cells: stimulates proliferation of erythropoietic cells and induces synthesis of hemoglobin. Although it would be simpler to identify erythropoietin as the Erythroid CSF, Metcalf (1981) points out that the action of erythropoietin does not explain some of the observations made on erythropoiesis in vivo. He therefore suggests the existence of a separate E-CSF, tentatively identified as a glycoprotein of molecular weight 24,000. (e) One should add to the list of hemopoietic growth factors the T cell growth factor, or interleukin-2 (Aarden 1979). I have already mentioned it in chapter 9, in proposing a model for the coordination of growth in size and cell DNA replication. Unstimulated T cells do not respond to interleukin-2, but when activated by antigens or mitogens they respond to it with proliferation in a dose-dependent manner. B lymphocytes do not respond to interleukin-2. Once activated, T lymphocytes can proliferate in the presence of interleukin-2 for indefinite periods of time. The production of interleukin-2 involves the interaction of activated lymphocytes with macrophages.

Other Growth Factors

The number of reported growth factors in the literature is limitless. To review all of them would require a separate volume. The growth factors discussed above should give a reasonable idea of the state of the art, of the kind of molecules we are dealing with, and of some of the mechanisms involved. For other growth factors, some of which (like hydrocortisone and estrogens) belong more to the field of classical endocrinology, the reader is referred to the literature on other growth factors and classic hormones (Sato and Ross 1979; Baserga 1981). The same comments must regretfully apply to the various adhesion factors that promote the interaction of a cell with its substrate.

INHIBITORY FACTORS

As I mentioned above, it is easier to identify growth stimulatory factors than inhibitory factors. The literature on growth inhibitory factors is actually abundant, but most of these putative inhibitors of cell proliferation have a very short scientific half-life and disappear quickly. The problem is that many factors can inhibit cell growth in vitro, but their biological significance is highly questionable. Perhaps I can best illustrate what I mean by an experience I had a few years ago, one that cannot be gleaned from the literature. A colleague of mine, a first-rate biochemist, had purified to homogeneity a liver protein that, when added to cells in culture, completely inhibited cell proliferation. His initial enthusiasm was somewhat tempered when he noticed that, after 3–4 days, the cells began to die. Since he was a good biochemist, he promptly found out that the liver protein in question was arginase, which simply deprived cells in culture of an essential amino acid. This protein had met our first criteria for an inhibitory factor, purity, but did not meet our other criteria, i.e., it should not be toxic, its effect should be reversible, and it should stop cells in specific stages of the cell cycle.

A growth inhibitory factor that meets all these criteria is the one described by Holley et al. (1983). It was isolated from medium conditioned by high density cultures of BSC-1 cells, epithelial cells of African green monkey kidney origin. It is a protein, is active on BSC-1 cells at ng/ml concentrations, and arrests them in the G_0 phase of the cell cycle. It also inhibits cells of lung and mammary origin, although some other cell lines are refractory to its inhibitory effect.

Although they have not been purified yet, we cannot avoid saying a few words on chalones, which have generated a considerable amount of literature. I will follow for this purpose the review of Iversen (1981), which is detailed, balanced, and easy to read.

Iversen (1981) defines chalones as follows: (1) They are naturally occurring, physiological inhibitors of cell proliferation; (2) they must be produced in and be present in the tissues on which they selectively act; (3) they are tissue- (or cell-line) specific, and their action is nontoxic and reversible. Iversen gives a long list of candidates for the role of chalones, including: (1) epidermal chalone, in which he distinguishes a G_1 chalone and a G_2 chalone, respectively arresting epidermal cells in G_1 and G_2; (2) a granulocyte chalone, that inhibits the proliferation of immature granulocytes; (3) an erythrocyte chalone; (4) several lymphocyte chalones; (5) a liver chalone; and (6) a number of chalones originating from tumor cell lines. Several other putative chalones from different tissues or sources have also been proposed.

The positive aspect of chalone research is the concept. A tissue specific inhibitor, produced by the same cells that are sensitive to it, fits in with our ideas about autocrine growth factors. The negative aspect is that, despite intense efforts, no chalone has thus far been purified. Clearly, until that is done, we can accept the concept only tentatively.

Chapter 11

Biochemical Events in the Cell Cycle

The first two parts of this book have described the tools used to study cell reproduction, while this last part really deals with the fundamental processes. Ultimately, our goal is to identify the growth factors and the genes and gene products that control the division of cells. In chapter 10 we have looked at the environmental signals, and in this and the next chapters, we will look at the gene products and the genes that are important for cell proliferation.

Before we look at the biochemical events in the cell cycle, though, we have to spend some time on a couple of semantic problems. The reason is that if we wish to appreciate some of these events in their proper perspective, we need to agree on a few definitions. Clearly, it is of little use to discuss G_1 events if one does not wish to assign a distinct G_1 period to the cell cycle. It is therefore important at this point that we should discuss a few questions that are often brought up in the literature.

DOES THE CELL CYCLE EXIST?

The answer to this question is: of course not. Once in a while, neophytes in the field of cell proliferation, rediscover this simple truth and they get all excited about it. Variations on the theme are the other two cosmic questions: does G_1 exist? and does G_0 exist? The answers are always the same: of course not.

The nomenclature of the cell cycle (G_0, G_1, S, G_2, and the cell cycle itself) is purely a scientific notation. Let me quote from the paper by Quastler and Sherman (1959): "Cells in mitosis are distinguished from intermitotic cells by direct observation; DNA labeling combined with high resolution autoradiography distinguishes cells synthesizing DNA from cells which are not. These two dichotomies divide the generative cycle into four phases: mitosis (M), presynthetic interval or post-mitotic gap (G_1), synthesis phase (S) and post-synthetic interphase or pre-mitotic gap (G_2) . . . *Ordinarily,* cells pass through them in sequence." There are only two facts here: mitosis and DNA synthesis, to which we can add the doubling of all cellular components. But some cells (in fact most cells after the first stages of embryonic development) have an interval between completion of mitosis and onset of DNA synthesis, and a second interval between completion of DNA synthesis and prophase. For simplicity, we call these intervals G_1 and G_2. Instead of saying that a cell has completed mitosis but has yet to begin DNA synthesis, we say a cell is in G_1, which is much shorter—the whole purpose of a scientific notation. True, after a while we get casual about our convenient notations, and we often refer to them as actual entities. It is then right that someone should remind us that our signs do not exist per se but are simply convenient notations. Thus G_1 does not exist, but what exists is, in many cells, a period of time between completion of mitosis and onset of DNA synthesis—just as S phase does not exist, but DNA synthesis does. The real question, of course, is not whether G_1 exists or not, but *why* so many cells, after completing mitosis, seem to hesitate for a few hours before they begin to synthesize DNA.

Let me make an analogy. The SV40 genome has 5,243 base pairs. Because it has been fully sequenced, the nucleotide residues have been numbered, from 1 to 5,243, beginning at the origin of replication. When we discuss SV40, we use nucleotide numbers, conveniently and casually, to localize functions and domains. Thus, we may say that the coding sequence for large T and small t begins at nucleotide residue 5,163. No one in his right mind would believe that these numbers actually exist, but how convenient they are! We can extend the analogy. The SV40 genome is conventionally represented as a circle in which the early mRNAs go counterclockwise and the late mRNAs clockwise, with the EcoR1 restriction site at 12 o'clock and the origin of replication at 8 o'clock. There is of course no compelling reason why the early or late mRNAs should go one direction rather than another. Similarly, there is no compelling reason why the diagram of the cell cycle should always be represented with cells moving clockwise, but I do not think that we should have a special symposium to inquire whether the cell cycle goes clockwise or counterclockwise. In fact, while the SV40 genome *is* circular, there is no reason to believe that the cell cycle is a perfect circle. Indeed, in some diagrams, it is represented with protuberances, like crown gall tumors, sticking out from G_1 and G_2. Let us not forget that these diagrams are simplified notations and that they may apply to some cell types but not to others, but let us also remember that they allow us to write shorter, clearer papers.

The question, Is G_0 different from G_1? was justified until a few years ago in the context of whether G_0 was a different state from G_1, or simply a long G_1. The answer to this question will be given below and it is that G_0 is qualitatively different from G_1 and, again for simplicity, one is justified in using the term G_0 for cells that are in an obviously different physiological state than cells in G_1. True, cells in late G_1 are probably in a different physiological state than cells in early G_1, and one could rightly insist that these two states (and many others) should be distinguished from each other. For the moment, differences in G_1 are still in a nebulous state (except at the G_1/S boundary), while G_0/G_1 differences are now clear-cut. As our knowledge grows, there is no reason why eventually the cell cycle should not be subdivided into additional sections characterized by specific events.

We are then justified to speak of G_1 events (or G_0 events) even though we know that some cells do not have a G_1 period and some cells are incapable of entering G_0. By this definition, G_1 events are simply biochemical events that have been described in cells between mitosis and S phase. We should always remember that some of these events may occur only in some cell types and not in others, and that, in some cell lines without G_1, they may occur before mitosis. What is then the usefulness of describing in detail G_1 (or G_0) biochemical events?

We have said before that it is important to investigate why so many cells, in vivo as well as in vitro, seem to hesitate, after mitosis, before entering the S phase. What makes them hesitate? Is there something they have to do, to complete, before cell DNA replication begins? What is it? This brings us to a second important semantic problem, i.e., what are the events that are just necessary for and those that actually control cell proliferation.

CONTROLLING EVENTS AND PREREQUISITE EVENTS

A partial list of cellular components that at one time or another have been announced as *controlling* cell proliferation would include: membrane glycoproteins, membrane glycolipids; ATP, cyclic AMP, cyclic GMP; cell size, nuclear size; synthesis of ribosomal RNA, amount of ribosomal RNA, number of ribosomes; phosphorylation of ribosomal proteins, size of deoxynucleotide pool, ribonucleotide reductase; DNA polymerase, phosphorylation of tyrosine; histone phosphorylation, histone dephosphorylation, histone acetylation, histone deacetylation; nonhistone chromosomal proteins (either synthesis or phosphorylation), plasminogen activator, Na^+ fluxes, intracellular pH, Ca^{2+}, Mg^{2+}; ornithine decarboxylase; calmodulin; p53, c-*myc*. This list is not intended to be a disparaging one. I have contributed to it myself, and more than once. In a broad sense, these cellular components control cell proliferation; in a strict sense, none of them does. It depends on our definition of what is a "control" of cell proliferation.

We have already seen that cell division requires growth factors and we shall see that it also requires a long series of biochemical events, all of which depend on a reasonably normal functioning of all metabolic processes of the cell. Let us leave aside the external growth factors that are easier to sort out, and let us look at cellular components only. If we wish, we can say that all the components listed above, all the biochemical events discussed in the previous chapters and in this one, *and* all metabolic processes of the cell control cell proliferation. This definition avoids controversy and makes many of us proud discoverers of what controls cell proliferation. But is it what we are really after?

Let us remember that in vivo as well as in vitro, cells can be quiescent under conditions that have to be considered physiological. The hepatocytes of healthy animals are living under normal physiological conditions and yet they do not proliferate. When a given stimulus is applied, most of them go into DNA synthesis and divide. Granted, the signal, i.e., growth factors (stimulatory or inhibitory), must come from the outside, and in this respect growth factors do control cell proliferation. But when we move inside the cell, we wish to know what is the trigger that sets in motion all these biochemical processes eventually leading to DNA synthesis and cell division. Very likely, there may be more than one trigger, but for simplicity let us assume there is only one. The question I am addressing in this section is: if the trigger is the intracellular process that controls cell proliferation, how do we define it? This is not just a philosophical question. It is a question that, if answered properly, will help a great deal in focusing our attention on the real problem.

Let us start with things like ATP and Mg^{2+}. These molecules are obviously needed for many metabolic processes of the cell. It is not surprising that when the intracellular concentration of ATP (or Mg^{2+}) falls below a certain level, cells stop dividing. Many other metabolic processes must come to a standstill, and the effect on cell proliferation is clearly aspecific. What one can say about molecules like ATP, Mg^{2+}, Ca^{2+}, etc., is that an optimal intracellular concentration is *necessary* for cell division, but they do not control cell division, because they are present in optimal amounts even in quiescent cells.

More complicated is the situation with certain enzymes, such as thymidine kinase, ribonucleotide reductase, DNA polymerase, and others. Such enzymes are absent or nondetectable in quiescent cells, and they show dramatic increases at the G_1/S boundary. It is tempting to consider them as controlling cell proliferation, and in a sense they do. Without DNA polymerase α, for instance, there would be no DNA synthesis. Without deoxynucleotides, there would be no DNA synthesis, either. And yet, quiescent cells can live very happily without an active DNA synthesizing machinery. Undoubtedly, deoxynucleotides and enzymes for DNA synthesis are both *necessary* and *specific* for cell proliferation. But do they really control cell proliferation?

Levels of thymidine kinase have often been thought to control the extent of cell proliferation. Since tk^- cells can happily grow in standard media and since certain species (squirrels, for example) do not have thymidine kinase at all (Adelstein, Lyman, and O'Brien 1964), we can rule that out without hesitation. The claim for ribonucleotide reductase is much more substantial. When this enzyme is inhibited, for instance, by hydroxyurea, DNA synthesis stops. It is true that, in regenerating liver, ribonucleotide reductase activity reaches a maximum at 50 hours after partial hepatectomy (Larsson 1969) when cell proliferation has all but ceased (Grisham 1962). That would not necessarily rule out a role of ribonucleotide reductase in the control of cell proliferation. The same comments apply to the whole DNA synthesizing machinery.

But it seems to me that by the time a cell reaches the G_1/S boundary, that cell is committed to at least DNA synthesis. I would like to define an element that controls cellular proliferation as that element (or group of elements), say a gene product, that *initiates* the chain of events leading to DNA synthesis and mitosis. An example of such a gene product, but unfortunately not of a cellular gene, is the SV40 T antigen. When SV40 T antigen, or the T antigen coding gene, are purified and microinjected into quiescent mammalian cells, they stimulate cell DNA synthesis, without any addition to the culture of nutrients or growth factors (Mueller, Graessmann, and Graessmann 1978; Tjian, Fey, and Graessmann 1978; Galanti et al. 1981). We can say that, in these cultures, T antigen determines or controls the extent of cell entry into S phase. I would like to define the cellular components that control cell proliferation as the cellular equivalents of SV40 T antigen. The p53 protein (see below) is a possible candidate, but we have no definite proof for it.

The distinction between the cellular equivalents of T antigen and the DNA synthesizing machinery is obviously narrow. Both are specific prerequisites for cell DNA synthesis. Perhaps, one difference could be their dependence on environmental signals. As we shall see in chapter 13, the mRNA for thymidine kinase can be induced by serum stimulation of quiescent cells. However, if one uses (as we did) a ts mutant that blocks in G_1 at the restrictive temperature, serum stimulation at the nonpermissive temperature fails to induce the tk-mRNA. One could say that the induction of tk-mRNA requires the expression of a previous cellular gene (or genes). Some serum-inducible genes, instead, will be induced even at the restrictive temperatures as if they were directly dependent on serum signals (chapter 13). I would like therefore to propose the following definitions: (1) components like DNA synthesizing enzymes *regulate* cell proliferation; (2) molecules like the SV40 T antigen *control* cell proliferation; and (3) molecules like ATP and Mg^{2+} are simply aspecific components of cell metabolic processes. These definitions are summarized in Table 11.1. In societal terms, the cellular equivalent of T antigen is the decision maker, while the DNA synthesizing machinery is the executor, the one who really says how and when things should be done but removed from deciding whether something should be done or not. Biologically, this distinction is more difficult to codify. I will try.

Suppose there are three gene products that, when present, induce a quiescent cell to enter S phase. Suppose also that these three hypo-

Table 11.1. Intracellular components and processes necessary for cell proliferation can be classified into three categories or levels of regulation.

Level	Description	Example
1	Necessary and specific for cell cycle progression. Absent or markedly decreased in nonproliferating cells. Under the control of environmental signals. Do not depend on the previous expression of other cell cycle genes (cdc genes).	"Start" gene in yeast; SV40 T antigen
2	Necessary and specific for cell cycle progression. Absent or markedly decreased in nonproliferating cells. *Not* under the direct control of environmental signals.	DNA synthesizing enzymes
3	Necessary but aspecific. Part of ordinary metabolic processes and present also in nonproliferating cells. Cells will stop growing in their absence.	ATP; Mg^{2+}; Ca^{2+}; RNA polymerase II

thetical gene products appear in direct response to an environmental signal and are not dependent on previous expression of other cellular genes. These gene products I would define as controlling cell proliferation. Enzymes for DNA synthesis do not qualify as controlling elements under this definition (see below). Incidentally, temperature per se has no effect on thymidine kinase activity, so its appearance must depend on previous cellular processes that do not occur when the cells are blocked in G_1. Cell fusion experiments indicate the same conclusion for the whole DNA synthesizing machinery. Thus in this definition, the controlling elements for cell proliferation are those that are a necessary prerequisite *and* are induced directly by the environmental signals or do not depend on previous expression of other cellular genes.

We can now look at the events that have been described in various phases of the cell cycle, with a view to evaluating their importance in the context of the question which is the truly fundamental question of this book: what controls cell proliferation?

G_1 AND G_0

We will first look at events that have been described in either G_1 cells or in stimulated G_0 cells or both and we will then sort out some of the characteristics that distinguish G_0 from G_1 cells. Since it is generally believed that the control of cell proliferation resides in G_1, it is not surprising that the list of events given above largely refers to events that occur in G_1. We will take some of these events in consideration in somewhat more detail.

Cell Size and Amount of Cellular RNA

A first group of events that actually represents a perfect illustration of our semantic problem includes those events related to the size of the cell, such as nuclear size, cell size, number of ribosomes, and synthesis or accumulation of rRNA. The relationship between growth in size and cell DNA replication has already been discussed in chapter 9. We need only reiterate here that: (1) there is ordinarily a strong correlation between cell size (measured as such, or as amount of RNA or proteins) and entry into S; but (2) cells can enter S phase with subnormal amounts of proteins or RNA; therefore cell size does not control cell DNA replication. However, many references in the literature (summarized in Baserga 1981) clearly indicate that proliferating cells have an increased rRNA metabolism, and that RNA amounts increase rapidly during G_1. The best example is illustrated in Fig. 4.2, which shows the progressive increase in the amount of cellular RNA occurring in cells going from one mitosis to the next one. We have also already mentioned that cell size may regulate mitosis, although here too there are exceptions (Zetterberg, Engstrom, and Larsson 1982). It should be mentioned at this point that RNA polymerase levels are also increased in proliferating cells. This is not so dramatic in cells in continuous cultures, but it is quite obvious in PHA-stimulated lymphocytes. The increase is illustrated in Table 11.2, which is taken from the paper by Jaehning, Stewart, and Roeder (1975). Note that the levels of all three RNA polymerases are increased, and that the levels increase promptly after PHA stimulation, even before DNA synthesis begins (24–48 hours after PHA).

Although the levels of the three RNA polymerases increase dramatically, RNA synthesis increases, but only modestly, during the transition from a resting to a growing stage. I must emphasize this point because some investigators, struck by the marked increase in cellular DNA synthesis occurring in serum-stimulated cells, look rather skeptically at the modest increases in RNA synthesis reported in the literature. However, one should remember that, in going from G_0 (or mitosis) to the end of S, a cell only doubles its RNA amount. Since cells, even G_0 cells, have a basal level of RNA synthesis, and since RNA amounts per cell increase linearly (Fig. 4.2), one expects,

Table 11.2. RNA polymerase levels in phytohemagglutinin-stimulated lymphocytes.

Days in culture	PHA	Units/10³ cells		
		I	II	III
0	—	135	191	23
0.5	+	151	213	23
1	+	433	477	67
2	+	1,013	663	148
3	+	1,768	953	350
4	+	2,241	1,502	406
4	—	77	141	12

Adapted from Jaehning, Stewart, and Roeder (1975).

at the most, a doubling in RNA synthesis. More than a doubling would require a concomitant increase in RNA degradation; but, if anything, RNA degradation decreases in cells stimulated to proliferate (see review by Baserga 1981). It is therefore logical that any stimulation of RNA synthesis would be modest. DNA synthesis is more impressive because it starts from virtually zero. If we took as the basal levels of RNA synthesis, cells in mitosis (where RNA synthesis is completely inhibited, see below), we would also find that RNA synthesis in G_1 is spectacularly increased. But in quiescent cells, RNA synthesis still goes on . . .

Intracellular Ions and Transport Across Membranes

A second group of events that has been related to G_1 progression or the exit from G_0 includes the intracellular concentrations of certain ions, and the transport of small molecular weight nutrients across the membranes. As stated by Pardee et al. (1978): "The rate of transport of small molecular weight nutrients across the cell membrane has been proposed as a primary regulator in switching cells between quiescence and proliferation." The review of Pardee et al. (1978) gives an excellent summary and many references on the increased transport of nutrients that occurs in cells stimulated to proliferate. Popular among the low molecular weight components has been Na^+. Within a few minutes after addition of serum to quiescent cell populations, there is a stimulation of the Na^+ K^+ pump leading to an increase entry of Na^+ (and K^+) into cells (Smith and Rozengurt 1978). This observation has been repeated many times (see for instance Koch and Leffert 1979). A careful study of intracellular Na^+ and K^+ during the cell cycle can be found in the paper by Mummery et al. (1981), which also contains many useful references. Their conclusion is interesting: "The transient increase in ($Na^+ - K^+$) ATPase mediated K^+ influx at the G_1/S transition is a prerequisite for entry into S phase, while maintenance of adequate levels of K^+ influx is necessary for normal rate of progression through the rest of the cell cycle." The reader should check this statement and the data of the paper by Mummery et al. (1981) against our definitions in Table 11.1. I would like to point out that in this case as well as in many other cases to be discussed below, there are two distinct components in the data given. The first component is the actual measurement of the intracellular concentration of K^+ (or Na^+ or Ca^{2+}, or whatever).

There is no reason to doubt the correctness of the data, which constitute a valuable addition to the general profile of what a cell does throughout the cell cycle. The second component is given by experiments in which a cell cycle block is produced by a drug that inhibits the accumulation of K^+ (or Na^+, etc.). It is the drug experiment that is used to estabish the "prerequisiteness," so to speak, of a cellular component. We have seen in chapter 8 the problems inherent to the specificity of drugs.

The importance of Ca^{2+} and Mg^{2+} in cell proliferation has already been alluded to above and in the previous chapter. There is no question that optimal concentrations of both Ca^{2+} and Mg^{2+} in the external environment are necessary for cell proliferation. Balk (1971) should be given the credit for the fundamental observation that optimal concentrations of Ca^{2+} are necessary for proliferation of normal fibroblasts, and a thoughtful study of both Ca^{2+} and Mg^{2+} can be found in the paper by Hazelton, Mitchell, and Tupper (1979). But we are interested here in intracellular concentrations. Rubin (1975) has claimed that intracellular Mg^{2+} concentration plays a central role in the proliferation response. As already mentioned above, the confusion here is about what is needed and what controls. This view is also supported by Hazelton, Mitchell, and Tupper (1979), who wrote: "In view of the fact that Mg is required for various aspects of macromolecular synthesis and transphosphorylation reactions, it seems quite reasonable that a continuous requirement for this ion would exist through G_1 phase because the processes themselves are continuous during this time." The relationship of intracellular levels of Ca^{2+} to cell proliferation can be best illustrated by the example of calmodulin, a major eukaryotic intracellular Ca^{2+} receptor, that regulates intracellular Ca^{2+} levels. The amount of calmodulin reaches a maximum in late G_1 and early S, and an anticalmodulin drug delays the progression of cells from G_1 into S (Chafouleas et al. 1982).

To conclude this section on ions and transport of small molecular weight nutrients, I want to quote directly from the review by Pardee et al. (1978): "No experiment has yet demonstrated that transport is a primary regulator of cell growth . . . A more likely possibility is that decreases in transport activity associated with quiescence are only feedback-type adaptations to a reduced requirement for nutrients."

Intracellular pH has also been proposed as a regulator of cell proliferation. However, careful studies on mitogen-stimulated lymphocytes have shown that the intracellular pH of cells in G_0, G_1, and G_2 is constant at about 7.2 and increases to 7.4 only during the S phase (Gerson and Kiefer 1983). I have already mentioned that an alkaline shock can stimulate cell proliferation in Swiss 3T3 cells. In my laboratory, we have microinjected into the nuclei of quiescent cells alkaline solutions, but we never observed any stimulation. In all fairness, the amount microinjected is by necessity small and may have not been sufficient to alter significantly the intracellular pH.

Cell Surface

Membrane changes almost invariably accompany the transition of cells from a resting to a growing stage. The changes in transport of low molecular weight nutrients just described reflect some of these changes. Baserga (1976) and Pardee et al. (1978) have already reviewed membrane changes related to cell cycle progression and cell proliferation. Among more recent findings, we should mention the role of cholesterol in cell growth. Cholesterol is a major structural component of the plasma membrane of animal cells. Chen (1984) has reviewed the relationship of cholesterol metabolism to cell cycle progression. His conclusions are: (1) quiescent cells synthesize little cholesterol; (2) cholesterol synthesis increases markedly in the G_1 phase; and (3) inhibition of cholesterol synthesis leads to inhibition of entry into S.

There are very important changes in plasma membranes related to the appearance of receptors for different growth factors, tyrosine phosphorylation, or the insertion of certain proteins (like the p21 protein of c-Ha-*ras*), but these are discussed in other sections of the book.

Polyamines and Ornithine Decarboxylase

Since ornithine decarboxylase (ODC) is a key enzyme in the production of polyamines, these two cellular components are treated together. There is no question that both ODC and polyamines are strongly correlated with cell proliferation. The reader is referred to the review by Heby and Jänne (1981), who marshalled the compelling evidence relating the increase in ODC activity and in cellular polyamine content to cell proliferation. In addition, a number of inhibitors of polyamine synthesis block cell proliferation. These are listed in Table 7 of the review by Heby and Jänne (1981) and include inhibitors of ODC (as for instance, DL-α-methylornithine) as well as inhibitors of another key enzyme, S-adenosylmethionine decarboxylase. The importance of polyamines in cell growth is underscored by the mutant C54 reported by Steglich and Scheffler (1982). This mutant from CHO cells is deficient in ODC (3 percent of the ODC activity of the parent cell line) and is auxotrophic for putrescine, requiring 10^{-5} M putrescine in the medium to maintain a normal growth rate. Spermine or spermidine can also serve as a polyamine source.

It is tempting to speculate that ODC and polyamines may be involved in regulating the growth in size of cells (and, indirectly, cell division). Thus, a unique nonhistone nucleolar protein of M_r 70,000, whose phosphorylation by a polyamine-dependent protein kinase yields a phosphoprotein that regulates rRNA transcription, was found to be identical, by several stringent criteria, with ODC (Atmar and Kuehn 1981). Indeed, Russell (1983) stimulated rRNA synthesis in *Xenopus* oocytes by microinjecting them with purified preparations of ODC. Along these lines, Durham and Lopez-Solis (1982), working with the isoproterenol-stimulated salivary glands of mice, reported that ODC activity correlated more with hypertrophy (cell growth in size) than with hyperplasia (cell division), and Kontula et al. (1984) found markedly increased levels of ODC mRNA in kidneys of mice treated with androgens, a treatment that produces cellular hypertrophy rather than hyperplasia.

If ODC activity and polyamines are related more to growth in size than to cell DNA replication, on the basis of the discussion in chapter 9 we could reconcile their correlation to cell proliferation with the seemingly contradictory findings that cell DNA synthesis can occur, in some cases, without any increase in ODC activity or polyamine content (Paul et al. 1978; O'Brien, Lewis, and Diamond 1979). Thus, infection of semipermissive cells by adenovirus 5, which induces cell DNA synthesis but not growth in size (see chapter 12), fails to induce ODC or S-adenosylmethionine decarboxylase or polyamine accumulation (Cheetham, Shaw, and Bellett 1982).

Finally, it should be remembered that different mechanisms may be operative in different cell types. While the irreversible inhibitor of ODC, DL-α-methylornithine, *usually* blocks cell proliferation, it has the opposite effect on bone marrow of mice, where it causes proliferation of erythropoietic precursor cells (Niskanen et al. 1983).

Synthesis of Specific Proteins

Terasima and Yasukawa (1966) were first to demonstrate a requirement for protein synthesis for the progression of cells through G_1. Since then, a copious literature has confirmed the requirement for protein synthesis for the transition of cells from M (or G_0) to S. An increased synthesis of nonhistone chromosomal proteins (histones will be considered separately) is characteristic of quiescent cells stimulated to proliferate (Tsuboi and Baserga 1972). We have already seen in chapter 9 that the amount of cellular protein also increases throughout the cell cycle, probably through both an increased synthesis and a decreased degradation. A third alternative is given by the very careful studies of Kruse, Miedema, and Carter (1967). These authors studied the utilization of amino acids in animal cells in a perfusion system where growth was sustained beyond confluence. They found that the rates of utilization of nutritionally essential amino acids was quite constant for cell densities ranging from 0.24 (preconfluent) to 8.4 (postconfluent) monolayers. For instance, for histidine, the rate remained constant at 3.3 ± 0.2 pmoles per hour per microgram of cell protein. However, a progressively increasing amount of synthesized protein appeared in the culture medium as the rate of proliferation decreased. It may be possible therefore that quiescent cells may keep their amount of protein constant not by decreased synthesis or increased degradation, but simply by increasing protein secretion. Of more interest than bulk proteins are the specific proteins that, at one time or another, have been reported to increase or to appear de novo in G_1 cells or in stimulated G_0 cells. Some of these proteins are listed in Table 11.3. I am excluding (arbitrarily) from this list the enzymes strictly connected with DNA replication, which will be treated under S phase, although some of them make their appearance late in G_1. The search for specific proteins in G_1 or G_0 is of course justified by the generally held belief that there resides the temporal control of cell proliferation. As one can see from Table 11.3, there are many claimants. Some of them may simply be red herrings, but one or two among them may

Table 11.3. Proteins whose synthesis is increased in G_1 or in quiescent cells after stimulation.

Protein	Cell type	Reference
41K	WI-38, human	Tsuboi and Baserga (1972)
MEP (35K)	3T3	Scher et al. (1983)
Calmodulin	CHO	Chafouleas et al. (1982; 1984)
R protein (labile)	A31 (3T3)	Rossow, Riddle, and Pardee (1979)
Several cytoplasmic proteins	3T3	Thomas, Thomas, and Luther (1981)
p53	Lymphocytes	Milner and Milner (1981)
Five proteins	3T3	Pledger, Howe, and Leaf (1982)
Actin	3T3	Riddle and Pardee (1980)
M_r 50,000	3T3	Gates and Friedkin (1978)
Ornithine decarboxylase	Several	Heby and Jänne (1981)

actually be *the* gene product(s) that everyone is looking for, the equivalent of the SV40 T antigen and of the yeast start gene, i.e., the gene product that initiates a new round of cell division.

We have aleady discussed above calmodulin and ODC. There are two other proteins that ought to be discussed in more detail: the R protein and the p53 protein. The R protein is a labile protein, with an estimated half life of 2 – 3 hours in untransformed cells, that is synthesized in G_1 and is necessary for the progression of cells from G_1 to S (Rossow, Riddle, and Pardee 1979). In transformed cells, the R protein is considerably more stable, with a half-life of at least 8 hours (Campisi et al. 1982). It has been proposed by Pardee and collaborators that the R protein, whose synthesis is sensitive to environmental conditions, needs to accumulate to a critical amount before a cell can pass the restriction point and proceed toward DNA synthesis. Perhaps related to this R protein is an anchorage-dependent step described by Campisi and Medrano (1983) in the G_1 of untransformed cells. This fleeting dependence on an anchoring substratum was much reduced in SV40-transformed cells, while chemically transformed cells gave intermediate results.

It should also be noted that a labile protein has been identified at the "start" point (see chapters 6 and 13) of the yeast *Saccharomyces cerevisiae* (Popolo and Alberghina 1984), although its role has not yet been defined.

The p53 protein is a transformation-related protein that is often found in higher amounts in transformed cells than in their normal untransformed counterparts (Linzer, Maltzman, and Levine 1979; Crawford et al. 1981). It has been detected in cells transformed by DNA viruses (Lane and Crawford 1979; Linzer and Levine 1979), by RNA viruses (Ruscetti and Scolnick 1983), by chemicals and X-rays (DeLeo et al. 1979), and in several human tumor cell lines (Crawford et al. 1981).The p53 protein is co-precipitated with the SV40 large T antigen by standard anti T antisera (Lane and Crawford 1979) and by monoclonal antibodies against T antigen (Gurney, Harrison, and Fenno 1980). This is because a substantial fraction of the cellular p53 is bound to large T antigen. However, a measurable fraction is not and can be precipitated, without T, by monoclonal antibodies against the p53 protein (Gurney, Harrison, and Fenno 1980; Mercer, Avignolo, and Baserga 1984). Although present in uninfected embryonal carcinoma cells (Linzer and Levine 1979), it is not detectable in several untransformed cell strains (Crawford et al. 1981). The amount of p53 protein varies greatly in different human cell lines from ≤ 0.2 μg/g of total protein in normal WI-38 cells to 36 μg in EBV-transformed Raji cells, 316 μg in cells transformed by adenovirus 5, to 450 μg (always per g of total protein) in SV80, human cells transformed by SV40 (Thomas et al. 1983). Rather puzzling is the finding of the same authors that the cellular levels of p53 are very low (≤ 0.4 μg/g total protein) in HeLa cells, which, by any standard, are vigorously growing cells. The synthesis of the p53 protein is markedly increased in mixed populations of lymphocytes stimulated by concanavalin A (Milner and Milner 1981). Its relationship to actively dividing cells has suggested to a number of investigators that the p53 protein may play a role in the regulation of cell proliferation (Linzer and Levine 1979; Milner and Milner (1981). A role for the p53 protein in cell proliferation is further suggested by the experiments already discussed in chapter 8, in which a monoclonal antibody against the p53 protein was microinjected into the nuclei of quiescent Swiss 3T3 cells, which were subsequently stimulated with serum (Table 8.3). Microinjection of the p53 antibody at the time of serum stimulation (± 2 hours) inhibited the entry of Swiss 3T3 cells into the S phase (Mercer et al. 1982). Further studies (Mercer, Avignolo, and Baserga 1984) showed that microinjection of a monoclonal antibody against the p53 protein did not inhibit the progression of cells through G_1. The authors concluded that the p53 protein may regulate the exit of cells from G_0 into the cell cycle (see chapter 9).

Histone Synthesis and Modifications

Despite an occasional report to the contrary, the synthesis of core histones occurs during the S phase; accordingly, it will be treated there. There are, however, some exceptions, and, in addition, some histone modifications occur in G_1.

Basal synthesis of histones in G_1 goes on at a much lower rate than S phase histone synthesis, about 1–2 percent, with the exception of H1 whose synthesis during G_1 occurs at a significant rate (Wu and Bonner 1981). Certain variants of histones, like H3.1 and H3.2, are synthesized exclusively during S phase; others are not synthesized in G_1 or G_2 but they are synthesized in G_0. The situation is summarized in Table 11.4, taken from the paper by Wu, Tsai, and Bonner (1982). These authors proposed that the observation that synthesis of H2A.1 and H2A.2 occurs in quiescent but not in G_1 or G_2 cells is a demonstration that G_0 is a qualitatively different stage from G_1.

Of several excellent papers on histone phosphorylation during the cell cycle, we will follow for simplicity the review by Gurley et al. (1978), which is truly comprehensive. H1 is phosphorylated during G_1 (Fig. 11.1), this type of phosphorylation being cumulative, i.e.,

Table 11.4. Histone variants synthesized in various states.[a]

S phase	Quiescent	Basal (G1 and G2)
H4	H4	H4
H2B	H2B	H2B
H2A.1	H2A.1	NS
H2A.2	H2A.2	NS
H2A.X	H2A.X	H2A.X
H2A.Z	H2A.Z	H2A.Z
H3.1	NS	NS
H3.2	NS	NS
H3.3	H3.3	H3.3

From Wu, Tsai, and Banner (1982).
a. NS: not synthesized.

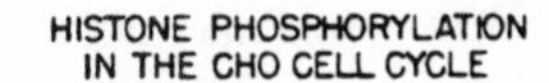

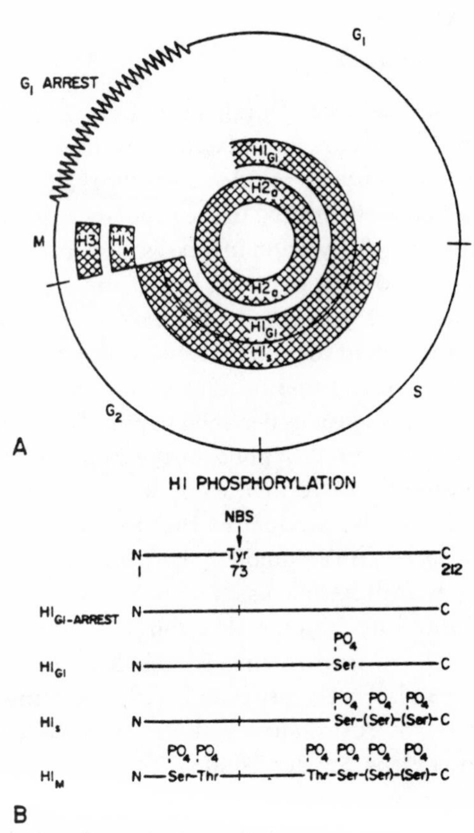

Fig. 11.1 Histone phosphorylation in the cell cycle of Chinese hamster cells. (A) Diagram of the cell cycle. The shaded areas indicate the phases of the cycle in which histones H1, H2A, and H3 are phosphorylated. (B) Diagram of the distribution of phosphorylation sites within the histone H1 molecule during the cell cycle. NBS is N-bromosuccimide, used to generate peptide fragments. (Reprinted, with permission, from Gurley et al. 1978.)

occurring on increasing numbers of H1 molecules as the cells progress through the cell cycle. The regional location of the G_1-phosphorylation within the H1 molecule was determined by bisecting the molecule with N-bromosuccinimide at tyrosine residue 73. G_1-phosphorylation occurs on a serine in the COOH-terminal portion of H1. This phosphorylation does not require unique copy gene transcription because it occurs normally at the restrictive temperature in tsAF8 cells (Pochron and Baserga 1979), which have, at 40°, a nonfunctional RNA polymerase II (see chapter 6).

Protein Phosphorylation

Apart from histones, other proteins are often phosphorylated during G_1 or when G_0 cells are stimulated to proliferate. For instance, nonhistone nuclear proteins are also phosphorylated in a cell-cycle dependent manner, as illustrated in the paper by Allfrey et al. (1974). Because protein phosphorylation increases in actively dividing cells, the theory that it somehow controls cell proliferation has been around for many years. It reached a peak of popularity when cyclic nucleotides were thought to hold the key to the control of cell proliferation, especially since some protein kinases were cAMP dependent. As the cyclic nucleotides declined in popularity as growth controls (see below) so did protein phosphorylation, but the theory was resurrected, more attractive than ever, when Collett and Erikson (1978) reported that the product of the transforming gene of the avian sarcoma virus (ASV), $pp60^{src}$, had protein kinase activity. Most of the cellular protein kinases were known to phosphorylate serine or threonine, but Hunter and Sefton (1980) demonstrated that the $pp60^{src}$ phosphorylates tyrosine. Protein kinases with specificity for tyrosine have also been associated with the transforming proteins of other retroviruses, such as Rous sarcoma virus, Abelson murine leukemia virus, Snyder Theilen strain feline sarcoma virus, and Y73 avian sarcoma. There are at least two growth factors that phosphorylate tyrosine: PDGF and EGF (see chapter 10). The report that the transforming protein of a primate sarcoma virus, p28 *sis*, and PDGF have extensive sequence similarity further supports the analogies between transforming proteins of retroviruses and growth factors, and underlines the importance of tyrosine phosphorylation in cell growth.

Another protein whose phosphorylation seems to be correlated with growth is the S6 protein of the 40S ribosomal subunit. When quiescent Swiss 3T3 cells are stimulated by serum, there is a very rapid and multiple phosphorylation of the S6 protein, which is essentially completed within 60 minutes after stimulation (Thomas et al. 1980). The phosphorylation is not inhibited by cycloheximide, indicating that the regulation is at a post-translational level. It should be noted, however, that protein S6 is rapidly phosphorylated (within 5 minutes) after exposure of 3T3 cells to insulin (Smith, Rubin, and Rosen 1980), which is not mitogenic for 3T3 but has often been correlated to growth in size of the cells (see, for instance, Zetterberg, Engstrom, and Larsson 1982). Indeed, Thomas et al. (1982) have confirmed that although serum, EGF, and insulin all cause phosphorylation of protein S6 in 3T3 cells, only serum and EGF stimulate DNA synthesis.

A broader picture is given by the studies of Chambard et al. (1983), who showed that factors that stimulate cell DNA synthesis, like serum, thrombin, or PDGF, phosphorylated both protein S6 and a cytosoluble peptide of 27,000 daltons in Chinese hamster cells, while insulin failed to phosphorylate the 27,000 daltons polypeptide.

We have already discussed above the nucleolar protein of M_r 70,000, which is phosphorylated by a polyamine-dependent protein kinase and is identical with ODC (Atmar and Kuehn 1981). Since this protein seems to regulate rRNA transcription, it probably has to do again with cell size. On the other hand, RNA polymerase II is phosphorylated by a cAMP independent nuclear protein kinase, called NII (Stetler and Rose 1982).

This very brief discussion clearly indicates that, in studying cell proliferation, one cannot ignore protein phosphorylation. The question is: how can we reconcile the requirement for unique copy gene transcription (already discussed in chapter 9) with protein phosphorylation, often regulated at a post-translational level? I would like to propose that protein phosphorylation, in general, correlates not with stimulation of DNA synthesis but with growth in size of the cell. Since growth in size of the cell can make the cell responsive to lower concentrations of growth factors for cell DNA replication (see chapter 9), protein phosphorylation could correlate well with cell proliferation. This interpretation is compatible with the data in the literature, and explains the exceptions. For instance, we have seen how ODC activity correlates with cell proliferation, although it does not always correlate with cell DNA synthesis. It explains also why no protein kinase activity can be associated with the transforming protein of certain DNA oncogenic viruses, like SV40. As we shall see in chapter 12, SV40, through the T antigen coding gene, provides a gene product, the T antigen, that causes both cell DNA synthesis and stimulation of rRNA genes, i.e., growth in size, by a mechanism that must be independent from phosphorylation.

Cyclic Nucleotides

We mentioned above that cAMP and to a lesser extent other cyclic nucleotides were at one time proposed as regulators of cell proliferation. Although their role in cell proliferation is now considered to be marginal, it is still worthwhile to discuss them briefly, because they constitute an excellent illustration of the problem of premature generalizations in assigning growth control functions to intracellular components. We can now say that while cyclic nucleotides play a very important role in a cell's life processes, they have little to do with cell proliferation. We will take as a prototype cAMP.

Three sets of observations were taken as indicating a major role of cAMP in cell proliferation: (1) addition of cAMP (or its derivatives) to populations of cells in culture caused inhibition of DNA synthesis and mitosis (Bombik and Burger 1973); (2) addition of cAMP to morphologically transformed Chinese hamster cells caused the cells to revert to a morphologically untransformed phenotype (Hsie and Puck 1971); and (3) intracellular levels of cAMP decreased when quiescent cells were stimulated by serum (Sheppard and Bannai 1974). While these observations were (and remain) correct, further studies have shown enough exceptions to (2) and (3) as to make a role of cAMP in cell proliferation rather problematic. For instance, the levels of cAMP (and cGMP) in mouse salivary gland increase considerably after isoproterenol, a drug that causes a marked stimulation of DNA synthesis and mitosis in the same tissue (Durham, Baserga, and Butcher 1974). Similar observations have been made in cells in culture. In addition, Steinberg, Wetters, and Coffino (1978) have isolated a mutant of S49 cells that has no cAMP-dependent protein kinase activity. Since this mutant grows normally in suspension, one has to conclude that cAMP-mediated reactions are not necessary for cell proliferation. Mutatis mutandis, the same story could be said of plasminogen activator.

Differences between G_0 and G_1

As already mentioned in this chapter, the separation of G_0 as a distinct entity from G_1 belongs to that group of cosmic questions that generate more papers than data. I have already reviewed (Baserga 1976) some of the differences between G_0 and G_1 cells, which include: (1) the prereplicative phase of stimulated G_0 cells is usually substantially longer than the G_1 of the same cells; (2) there are differences in the sensitivity to inhibitors, such as L-asparaginase and L-histidinol; (3) there are differences in the complement of ribosomes, in the amount of cellular RNA (see Table 3.1), and in the electrophoretic patterns of nonhistone nuclear proteins and other proteins (see above); (4) most transformed cells, in particular virally transformed cells, cannot enter G_0. Epifanova (1977) has also reviewed in detail the differential sensitivity of G_0 and G_1 cells to external factors. All these differences, to which many more have been added since 1976, have been criticized because (it has been said) they are purely quantitative. I have never been able to understand why a quantitative difference is not an acceptable difference. The known differences between G_2 and G_1 cells are also purely quantitative, but it still is convenient (and fruitful) to keep G_2 separate from G_1. Personally, I have found the behavior of G_0 cells sufficiently different from that of G_1 cells that I find it more convenient to use the two terms to indicate two different physiological states.

This much said, let me add six more pieces of evidence, which seem to make the distinction more acceptable, this time on a qualitative basis: (1) the p53 protein is needed for the exit from G_0 but not for progression through G_1 (see above); (2) certain histone variants are synthesized in quiescent cells but not in G_1 cells (also discussed above); (3) using somatic cell hybridization, Rao and Smith (1981) showed that G_0 cells differ from G_1 cells with regard to their effects on the cell cycle progression of G_2 cells after fusion; unlike G_1 cells, G_0 cells upon fusion with G_2 cells are not able to inhibit the progression of the G_2 nucleus into mitosis; (4) the number of S_1 nuclease-sensitive sites in purified DNA, which is 0 in G_0 cells, increases to $22-43/10^5$ base pairs 6 hours after serum stimulation when the cells are still in G_1 (Collins, Glock, and Chu 1982); (5) the levels of calmodulin and of calmodulin mRNA are higher in G_0 than in G_1 (Chafouleas et al. 1984); and (6) Ide, Ninomiya, and Ishibashi (1984) have isolated a ts mutant of rat 3Y1 cells that grows normally in the exponential phase at the restrictive temperature but fails to enter S phase when stimulated from quiescence at the nonpermissive temperature. This mutant has the characteristics of a G_0 mutant, i.e., it is not ts for G_1, but it is ts for the transition from quiescence to S phase. Conversely, the mutant tsD123 described by Zaitsu and Kimura (1984) is ts in G_1 but not when stimulated from a G_0 state.

Finally, in chapter 13, the discussion on cell division cycle genes should put an end to this controversy that has already been going on far too long.

S PHASE

While it is possible that a low level of DNA synthesis may take place during G_1, there is no question that the bulk of DNA synthesis occurs in a discrete period of time of the interphase, which we call the S phase. Indeed, DNA synthesis and mitosis are the two markers of cell proliferation everybody agrees upon. Some cells may not have a G_1, others may not have a G_2, but they all have to synthesize DNA and undergo mitosis. Growth in size is the third obligatory component of what we could define as the basic aspects of the cell cycle. I shall discuss first DNA replication, then the enzymes connected with it and finally some of the other biochemical events (for instance, histone synthesis) that occur during the S phase.

DNA Replication

Most of this discussion is based on the review by Pardee et al. (1978). In eukaryotic organisms, each DNA fiber is divided into many replicating units, called replicons. Each replicon has a center, the origin, from which growing forks extend outward in both directions. Forks of adjacent replicons fuse with each other. By using a cell line in which the gene for dihydrofolate reductase was amplified 500-fold and Southern blots of labeled genomic digests, Heintz and Hamlin (1982) showed that the initiation of DNA synthesis in the chromosomes of mammalian cells is sequence specific. As in prokaryotes, an RNA primer is necessary for the initiation of DNA synthesis. Tseng et al. (1979) state that in mammalian cells, the synthesis of nascent DNA is primed by a nonaribonucleotide with nonunique sequences. Reichard and Eliasson (1979), studying polyoma DNA replication, found that the primer RNA, which they call initiator RNA, is synthesized by a polymerase which is a mammalian counterpart to primase for *E. coli.* An oligoribonucleotide of six to nine bases covalently attached to the 5′ ends of nascent DNA chains has also been described in SV40 (Hay and DePamphilis 1982). Once the cell has selected a locus for DNA synthesis, each DNA strand is replicated in both directions at a rate of $0.5-2\ \mu m/min^{-1}$. The spaces between the origins of replication vary in size from 7 to 100 μm.

In the very early stages of embryo development, DNA synthesis must occur almost simultaneously in all replicons, since the S phase is very short (see chapter 3). But in the cells of the adult animal, the S phase lasts 8 hours or more, and it has been known for a long time that initiation of DNA synthesis occurs at different times in different chromosomes and, furthermore, that a definite temporal order exists for DNA replication in different chromosomal segments (Taylor 1960). The inactive X chromosome of female cells, that does not begin replication until all other chromosomes are replicated, is an extreme example (Moorehead and Defendi 1963), but a reproducible temporal order can be observed in each chromosome. In fact, it has been demonstrated that the same DNA sequences replicate during the same intervals of the S phase from one cell generation to the next (Kajiwara and Mueller 1964; Mariani and Schimke 1984).

It remains a mystery how this temporal order is maintained, considering that the origin of replication is sequence-specific and that the DNA synthesizing machinery is presumably available throughout the nucleus. Since the inactive X chromosome is hypermethylated, I have often played with the hypothesis that DNA methylation may have something to do with the temporal order of DNA replication rather than with the regulation of transcription. But I cannot think of an intelligent experiment along these lines and, besides, the dogma is that DNA methylation regulates gene expression, and one should never fight dogmas. Openly, that is.

DNA Synthesizing Enzymes

We are not concerned here with the biochemical properties of these enzymes, but with their relationship to the cell cycle. Pardee et al. (1978) call the whole group of these enzymes the DNA polymerases and ligases, to which I would like to add at this point other enzymes connected with DNA metabolism, such as thymidine kinase, ribonucleotide reductase, and others.

A reasonable generalization is that most, if not all, of these enzymes, increase markedly in activity at the G_1/S boundary. A typical example is given by thymidine kinase. Thymidine kinase levels begin to rise at the onset of S phase; however, they persist at very high levels during G_2 and mitosis, (Stubblefield and Murphree 1968), i.e., even after the enzyme is no longer needed. This happens also with other enzymes. For instance, in regenerating rat liver, ribonucleotide reductase activity reaches a maximum 50 hours after partial hepatectomy (Larsson 1969) when DNA synthesis has all but ceased. Another point that should be made is that the activities of these enzymes can often be detected, albeit at very low levels, even in quiescent cells. It raises the question, still unresolved, whether the basal levels are or are not sufficient for DNA synthesis, the marked increases detectable at the G_1/S boundary being of modest physiological significance. Certainly, there is no special need for thymidine kinase: not only tk^- cells grow vigorously in culture, but some rodent species lack thymidine kinase altogether (Adelstein, Lyman, and O'Brien 1964).

Some of the enzymes whose activities are markedly increased at the G_1/S boundary are listed in Table 11.5. For each of them, I have given one illustrative example (PHA-stimulated lymphocytes, isoproterenol-stimulated salivary gland, etc.), but other examples can be found in the literature (see, for instance, the review by Pardee et al. 1978).

The activity of ribonucleotide reductase is regulated in an interesting way. The enzyme, a key one in the production of the deoxyribonucleotides required for DNA synthesis, is composed of two subunits, M1 and M2, which are inactive alone but are fully active when combined. Protein M2 increases in amount at the G_1/S boundary, but protein M1 is present in similar amounts throughout the cell cycle (Eriksson and Martin 1981).

Pardee and co-workers (Noguchi, Reddy, and Pardee 1983) have

Table 11.5. The DNA synthesizing machinery.

Enzyme	Example	Reference
Thymidine kinase	Fibroblasts in culture	Stubblefield and Murphree (1968)
Thymidylate kinase	HeLa cells	Brent, Butler, and Crathorn (1965)
Ribonucleotide reductase	Regenerating liver	Larsson (1969)
Dihydrofolate reductase	3T6 mouse cells	Johnson, Fuhrman, and Wiedemann (1978)
Deoxythymidylate synthetase	Salivary gland after isoproterenol	Pegoraro and Basergá (1970)
DNA polymerase	Lymphocytes	Bertazzoni et al. (1976)
Replitase	Chinese hamster cells	Noguchi, Reddy, and Pardee (1983)

found that many of these enzymes are associated in a multienzyme complex, which they call replitase, about the size of a mammalian ribosome. The replitase complex can be obtained from the nuclei of S phase cells but not of G_1 cells. It contains ribonucleotide reductase, thymidine kinase, thymidylate synthetase, DNA polymerase, dihydrofolate reductase, ATP-dependent topoisomerase, nascent DNA, and DNA template. It rapidly channels ribonucleoside diphosphates through many enzyme-catalyzed steps into DNA, while excluding the freely diffusible pool of deoxyribonucleoside triphosphates. In the words of the authors, the replitase complex "provides a high concentration of precursors for DNA synthesis at the replication site, microenvironmentally different from its surroundings." The demonstration of the replitase complex and the fact that the activities of its component enzymes increase coordinately at the G_1/S boundary raise the legitimate questions whether all these enzymes are regulated at a single locus. This may be so, but each of these enzymes probably has its own promoter because, by transfecting the DNA for human thymidylate synthase into mouse cells, Ayusawa et al. (1983) found that the enzyme was expressed by the transfected DNA and did not associate with the multienzyme complex.

In chapter 4, we saw that the concentrations of ribonucleoside triphosphates doubled from G_1 to S. The concentrations of deoxyribonucleoside triphosphates increase much more dramatically. Expressed in pmoles per 10^6 cells, and in CHO cells, dATP increases from 5.9 in G_1 to 61 in S phase, dGTP from 5.1 to 22, dTTP from 12 to 84, and dCTP from 20 to 310 (Hordern and Henderson 1982). Note also that the concentrations of ribonucleoside triphosphates are in nmoles, those of deoxyribonucleoside triphosphate in pmoles. Another interesting aspect of deoxynucleoside metabolism is that thymidine is phosphorylated in G_1 as well as in S phase, but in G_1 it is broken down and eliminated, while in S phase it is incorporated into DNA (Schaer and Maurer 1982).

Other S Phase Events

We have already mentioned that apart from basal histone synthesis, the bulk of histone synthesis occurs during the S phase. The evidence has been reviewed in the paper by Stein et al. (1982), which also gives the evidence that "histone mRNA sequences are present in significant amounts in the nucleus and cytoplasm of HeLa cells only during S phase, when histone protein synthesis occurs." This applies at least to the core histones, H4, H3, H2A, and H2B, whose mRNAs were detected by using as probes cloned genomic human histone sequences representing different histone gene clusters. The representation of histone mRNA sequences in G_1 compared with S phase cells, either in the nucleus or the cytoplasm, was about 1 percent. Furthermore, incorporation of [^{3}H]-uridine into core histone mRNA was essentially confined to the S phase (Plumb, Stein, and Stein 1983). In ts mutants of the cell cycle, blocked in G_1 at the restrictive temperature (see chapter 7), no histone synthesis can be detected (Pochron and Baserga 1979).

In addition to histone synthesis, histone phosphorylation continues through S phase, H1 being phosphorylated at two additional serine sites, different from the serine site phosphorylated during G_1 (Fig. 11.1). It will be noted from that figure that H2A is phosphorylated throughout the cell cycle (Gurley et al. 1978).

Another interesting observation concerns the appearance of the receptor for transferrin, which, as mentioned in chapter 10, is an obligatory component for the growth of cells in serum-free media. The best illustration is offered by the paper of Neckers and Cossman (1983), who used T lymphocytes stimulated by PHA. Exposure to PHA alone was sufficient to induce the appearance of receptors for interleukin-2 (IL-2). These receptors in turn were necessary for the induction of transferrin receptors. Antibodies to IL-2 receptors inhibited the entry into S phase, but only if the cells were exposed to the antibody before the appearance of the transferrin receptors. Exposure of cells to antibodies to transferrin receptors stops cells in S phase (Trowbridge and Lopez 1982). It almost seems as if the progression of cells through the cell cycle is regulated by the successive appearance of surface receptors for different growth factors, as already suggested in chapter 9.

Finally, we have to consider the well-established relationship between inhibition of protein synthesis and inhibition of DNA synthesis. It has been known for more than 20 years that inhibition of protein synthesis in cells in S phase causes a prompt decline in DNA synthesis. This observation was actually taken as a demonstration that histone synthesis and DNA synthesis could not be uncoupled (see Stein et al. 1982). A very good study on the subject has been carried out by Stimac, Housman, and Huberman (1977), and I am taking their paper for the basis of my comments. These authors used several methods to inhibit protein synthesis: cycloheximide, puromycin, emetine, pactamycin, amino acid analogs, and others. They all caused prompt reduction of DNA synthesis to plateau levels that were roughly the same as the inhibited rate of protein synthesis. By using DNA fiber autoradiography, Stimac, Housman, and Huberman (1977) found that within 15 minutes after inhibiting protein synthesis, the rate of movement at every replication fork was reduced. After 60 minutes, there was also a decline in the frequency of initiation of new replicons (see, however, the mutant described by Roufa, 1978, in which, at the restrictive temperature, protein synthesis is inhibited witout a concomitant inhibition of DNA synthesis).

G_2 AND MITOSIS

The G_2 Period

The first suggestion that gene expression was necessary for the progression of cells from S phase to mitosis can be found in the paper by Kishimoto and Lieberman (1964), who showed that when S phase cells were exposed to actinomycin D (0.33 μg/ml) or puromycin (5.0 μg/ml), they were blocked in G_2. Since then, numerous reports have confirmed that inhibitors of RNA and protein synthesis arrest the progression of cells through G_2. These studies have been summarized in an excellent review by Tobey, Petersen, and Anderson (1971). Using various inhibitors of protein synthesis (puromycin, cycloheximide, or infection with mengovirus), they related that the time of synthesis of the last proteins necessary for entry into mitosis varied between 10 and 120 minutes before mitosis depending on the cell line. With actinomycin D, the time of synthesis of the last RNA species essential for the entry into mitosis would vary between 30 minutes in Chinese hamster cells and 5–6 hours in Ehrlich ascites tumor cells. They also pointed out that there was no requirement for ribosomal RNA synthesis in G_2. Despite our reservations on the use of drugs, I think we can agree with Tobey, Petersen, and Anderson (1971) that "G_2 is a period of intense biochemical activity, which, once initiated, gives rise to a series of reactions . . . leading to cell division." That is, once DNA replication has been completed, the cell also has to decide to complete G_2 (and mitosis). That the progression of cells from S phase to mitosis is not automatic, is shown by the phenomenon of G_2 arrest. By looking at Table 1.2, one can see that G_2 (plus mitosis) can vary from 1.2 hours in rat duodenal crypt cells to 17.5 hours in KB cells. Since mitoses last from 30 minutes to 2 hours, most of the times given in Table 1.2 are a good indication of the length of G_2. However, there are not only cell lines with long G_2, there are also cell types that simply arrest in G_2. Credit for discovering that certain cells arrest in G_2 must go to Gelfant, to whose review (1977) the reader is referred.

The existence of cells blocked in G_2 can easily be ascertained by noticing that in some cases, after an appropriate stimulus, certain cells enter mitosis within a very short period of time, or by detecting unlabeled mitoses in populations exposed continuously for long periods of time to [^{3}H]-thymidine. Gelfant (1977) gives a long list of G_2-arrested cells in the epithelial cells of the epidermis, the kidney, the liver, certain tumors, germinating plant seeds, etc. G_2 arrest occurs also, under certain conditions, in cells in culture. The existence of a G_2 arrest and the use of drugs therefore suggest that some genes and gene products are necessary for the progression of cells through G_2. Indeed, Tobey, Petersen, and Anderson (1971) made a gallant attempt, in their review, to characterize division-related RNA species and proteins, but the methodologies available at that time did not allow more than a few hints. The progress, since then, has not been spectacular, but some new information deserves to be discussed.

First of all, there are the G_2-specific ts mutants (listed in Table 6.1) which, by themselves, indicate that some gene products are necessary for the orderly progression of cells through G_2. Very little is known about the ts defect in tsBN75 (Nishimoto, Takahashi, and Basilico 1980), but in ts85 cells there is a defect in H1 histone phosphorylation that prevents chromosome condensation and causes arrest in G_2 (Yasuda et al. 1981). According to Yasuda et al., however, all three phenomena may be dependent on the disappearance, at the nonpermissive temperature, of the basic chromatin protein A24. The defect is also known in 422E cells. At the restrictive temperature, 422E (derived from BHK21 cells) synthesize rRNA precursors normally but fail to produce 28S rRNA and large ribosomal subunits (Toniolo, Meiss, and Basilico 1973). When these cells are made quiescent by serum deprivation and are subsequently stimulated with serum, they enter DNA synthesis even at the restrictive temperature. Cell division, however, is impaired and the number of cells does not increase (Mora, Darzynkiewicz, and Baserga 1980). It seems therefore that the number of ribosomes, though not critical for entry into S, is critical for the completion of G_2 and mitosis.

We have already mentioned in chapter 7 that, in the classic fusion experiments reported by Rao and Johnson (1970), the G_2 cell follows certain rules that are clearly distinct from those of G_1 and S phase cells, namely: (1) a G_2 cell does not inhibit a G_1 nucleus from entering S phase; (2) a G_2 cell does not inhibit ongoing DNA synthesis in an S phase nucleus; and (3) an S phase cell cannot initiate a second round of DNA replication in G_2 nuclei. Rao and Johnson (1970) concluded, correctly "that there is a positive control of DNA synthesis in HeLa cells with appearance of inducer substances at the start of the S phase." However, the failure of G_2 nuclei to initiate a second round of DNA replication when fused with S phase cells indicates also a requirement for certain nondiffusable conditions intrinsic to the nucleus. In their experiments, Rao and Johnson (1970) noticed that in heterophasic S/G_2 cells the G_2 nuclei were delayed from entering mitosis by the presence of the S component. They suggested that this could be due to a dilution (by the S phase component) of the concentration of mitotic inducer substances present in the G_2 cells. Subsequently, Al-Bader, Orengo, and Rao (1978), using two-dimensional gel electrophoresis, found in HeLa cells a number of G_2-specific proteins necessary for the progression of cells from G_2 to mitosis.

Finally, it should be mentioned that, in HeLa cells, the levels of poly(ADP-ribosyl)ation and the activity of poly(ADP-ribose) polymerase reach a maximum during the G_2 period (Tanuma and Kanai 1982).

Mitosis

We are concerned in this section with the biochemical events that accompany mitosis, and not with the mitotic process itself. From a morphological point of view, a description of mitosis belongs in a textbook of general biology. However, the reader should consult the beautiful review by Pickett-Heaps, Tippit, and Porter (1982) in which the authors discuss, in a conceptually modern way, the mechanisms of spindle assembly and mitotic chromosomal motion.

Certain biochemical events characteristic of the mitotic cells have been known for several years and are discussed by Tobey, Petersen, and Anderson (1971) and by Baserga (1976). The reader is referred to these two sources for the original references. In the first place, there are striking membrane changes that make the surface of a normal mitotic cell resemble the surface of a transformed cell in interphase. The mitotic cell rounds up and is much less firmly attached to the supporting surface than the interphase cell. This is actually the basis for the mitotic selection procedure, which remains the best way to synchronize cells in culture (see chapter 5). Because mitotic cells are so loosely attached to the surface, a modest degree of shaking detaches them, while interphase cells are not dislodged. A similar phenomenon actually occurs in vivo. The cells lining the crypts of the small intestine constitute the proliferative compartment of the epithelial cell population of the intestinal mucosa. DNA synthesis and mitosis occur only in the crypt cells that are resting, as a monolayer, on a basement membrane. When a crypt cell is in mitosis, it almost completely loses contact with the basement membrane, and, in histological sections, mitotic cells stand out because of the proximity of their nuclei to the lumen of the crypts (Fig. 2.3).

Protein synthesis is markedly reduced during mitosis, to a level that is about 20 percent of the rate observed in interphase cells. Notable exceptions are nonhistone chromosomal proteins, whose synthesis continues during mitosis at the same rate as in interphase. Certain viruses, like poliovirus grow normally in metaphase-arrested cells, so that, clearly, proteins can be made during mitosis. It still remains a mystery why the synthesis of certain proteins and not others is inhibited.

A little less mysterious is the inhibition of RNA synthesis that occurs during mitosis. In the first place, the inhibition of RNA synthesis is total and it affects rRNA as well as mRNA and tRNA. The inhibition of RNA synthesis is not due to the lack of the appropriate enzymes. Benecke and Seifart (1975) analyzed the RNA polymerase activities solubilized from HeLa cells. With respect to amounts or activities of all three RNA polymerases, no significant differences were detectable between exponentially growing, mitotic, and middle-S phase cells. The defect seems to reside in the chromatin structure or in chromosomal proteins, since chromatin from mitotic cells has a decreased template activity in respect to chromatin from interphase cells, while the respective DNA are indistinguishable. In some respect, it is surprising that mitotic cells have not been used more to study factors that regulate gene expession. This may be largely due to the fact that the usual mitotic agents, like colchicine and colcemid, are quite toxic and cannot be used on cells for more than 4–6 hours. Perhaps the situation may change with the use of the reversible microtubule inhibitor Nocodazole, which has been described by Zieve et al. (1980). With this compound, HeLa cells continue to accumulate in mitosis at a linear rate, until 90 percent of the cells are mitotic. This inhibition is reversible, at least up to a certain point. Using Nocadozole, Mercer et al. (1984) have been able to demonstrate that the drug indeed causes a depletion in the amount of cellular RNA.

Finally, during mitosis, H1 histone is further phosphorylated, this time at serine and threonine residues of the NH_2-terminal half, while H3 is phosphorylated for the first time during the cell cycle (Fig. 11.1).

Chapter 12

Virally-Coded Proteins and Cell Proliferation

It is well established that certain oncogenic viruses can induce in quiescent animal cells cellular DNA synthesis and, under appropriate circumstances, mitosis. This stimulation of cell proliferation is limited to one or, at the most, two rounds of mitoses, unlike transformation, which permanently changes the phenotype. Since one of the characteristics of transformation is an increased cellular proliferation (see chapter 3), one might reasonably assume some relationship between the ability of a virus to induce cell DNA synthesis and its ability to cause transformation. There is, however, a clear-cut separation between the two processes. For instance, infection with

DNA oncogenic viruses can induce both cell DNA synthesis and transformation, but infection with certain RNA viruses, although causing transformation, does not result in a detectable transient increase in cellular DNA synthesis. With SV40 and polyoma almost 100 percent of the infected quiescent cells can be induced to enter S phase, but the frequency of transformation is much lower, around 1 in 10^5. Certain deletion mutants of SV40 can induce cell DNA synthesis but are incapable of permanently transforming cells (Soprano et al. 1983). A similar situation occurs with polyoma, in which certain mutants can trigger the mitogenic response but fail to induce permanent transformation in rat fibroblasts (Schlegel and Benjamin 1978).

In this chapter, we will take into consideration only the prompt and transient induction of cellular DNA synthesis and mitosis, which we should distinguish from the more permanent alteration of cell proliferation caused by transformation. The reader who wishes to have more details or to know the history of this topic should consult the several reviews in the literature, including those in Tooze (1980) and the one by Weil (1978). Table 12.1 gives a list of the DNA oncogenic viruses that cause prompt and transient induction of cellular DNA synthesis, with the original references.

The transient induction of cellular DNA synthesis by DNA oncogenic viruses has been the subject of extensive investigations, on the tacit assumption that these viruses stimulate DNA synthesis and mitosis by a mechanism analogous to that of growth factors. Because

Table 12.1. DNA oncogenic viruses that induce cellular DNA synthesis in quiescent animal cells.

Virus	Cell line or strain	Reference
Polyoma	Primary mouse kidney	Dulbecco, Hartwell, and Vogt (1965)
SV40	WI-38 human fibroblasts	Sauer and Defendi (1966)
Adenovirus (2 or 12)	Human embryonic kidney	Ledinko (1967)
Epstein-Barr virus	Human B lymphocytes	Robinson and Smith (1981)
Papilloma virus	Human HEK	Butel (1972)
Cytomegalovirus	Human embryonic cells	St. Jeor et al. (1974)
Herpes simplex virus 2	Rat hepatocytes	Isom (1980)

the genome of these viruses is much smaller than the genome of mammalian cells (5,243 base pairs in the SV40 genome against 3 × 10^9 base pairs in the mammalian genome), the identification of the viral gene product stimulating cell proliferation should be (and in fact was) much easier than the identification of the corresponding cellular gene. In other words, by identifying the viral gene product inducing a mitogenic response, one could hope to obtain a lead to the identification of a cellular protein functionally equivalent to the viral product. This belief has been forcefully stated by, among others, Weil (1978): "Studies on the virus-induced mitogenic reaction will contribute to the understanding of the molecular mechanisms involved in mitotic control and may ultimately also provide insight into the nature of various proliferative disease."

Alternatively, the DNA viruses may induce cell DNA synthesis by a different mechanism than growth factors. Even so, they would be of interest because, despite the differences, they could still have some common pathways with cellular genes involved in the control of cell proliferation. In this chapter, we will first examine the various DNA viruses that induce cell DNA synthesis and mitosis and identify the viral genes and gene products responsible for this effect. In the last section of the chapter, we will discuss the similarities and the differences between viral-induced and serum-induced cellular DNA synthesis.

SV40 AND POLYOMA

These two viruses will be treated together because their biological effects on quiescent animal cells are very similar. We will use SV40 as the prototype. Details on the polyoma-induced mitogenic response can be found in the review by Weil (1978) and in Türler (1980).

Induction of Cellular DNA Synthesis and Mitosis

Sauer and Defendi (1966) were the first to report that infection with SV40 stimulated cellular DNA synthesis in quiescent human diploid fibroblasts. Since then, many investigators have reported SV40-induced stimulation of cell DNA synthesis in a variety of quiescent cells, including 3T3 (for a review, see Tooze 1980). The stimulation of cellular DNA synthesis is accompanied by many of the phenomena that characterize serum-stimulated DNA synthesis, including a marked increase in the specific activity of enzymes involved in DNA synthesis, like thymidine kinase, DNA polymerase, dCMP deaminase, etc. (Kit et al. 1966). The synthesis of nonhistone chromosomal proteins is also stimulated (Rovera, Baserga, and Defendi 1972), just as in serum-stimulated cells (see above in chapter 11). Under appropriate conditions, SV40 infection induces cell division in quiescent 3T3 cells (Smith, Scher, and Todaro 1971). At high multiplicity of infection, the quiescent cells can even undergo several rounds of mitoses before returning to a nonproliferating stage. We will discuss later the ability of SV40 to induce growth in size and whether the mechanism of SV40-induced cellular DNA synthesis differs from that of serum stimulation. We are concerned here with identifying the gene in the SV40 genome that is responsible for the stimulation of cell DNA synthesis.

By using deletion mutants of SV40, Scott, Brockman, and Nathans (1976) showed that stimulation of cellular DNA synthesis depended on the functional integrity of the early region of the SV40 genome, more precisely the T antigen coding gene, also called the A gene. Fig. 12.1 shows a functional map of the SV40 genome. VP1, VP2, and VP3 are capsid proteins and there is clear evidence that they are not necessary for induction of cellular DNA synthesis. The two proteins that interest us in this chapter are the large T antigen (80,000 M.W.) and the small t antigen (17,000 M.W.), both encoded by the same gene, which extends from nucleotide residue 5236 counterclockwise to nucleotide residue 2587. When the transcript undergoes a 66 bases splicing, the gene produces a large mRNA which, however, has a termination codon at nucleotide 4641 and is translated into the small t antigen. A larger splicing (346 bases) eliminates the first termination codon, and produces a smaller mRNA with a termination codon at nucleotide residue 2693. This mRNA is translated into the large T antigen, which is easily detectable by indirect immunofluorescence or by gel electrophoresis of immunoprecipitates. The large T antigen is mostly located in the nucleus, while the small t is largely cytoplasmic.

In 1976 Graessmann and Graessmann, using *E. coli* RNA polymerase, transcribed in vitro a cRNA from the A gene of SV40. They then microinjected this cRNA directly into the nuclei of quiescent mammalian cells. A large fraction of the microinjected cells became positive for the SV40 T antigen and entered into S phase. Subse-

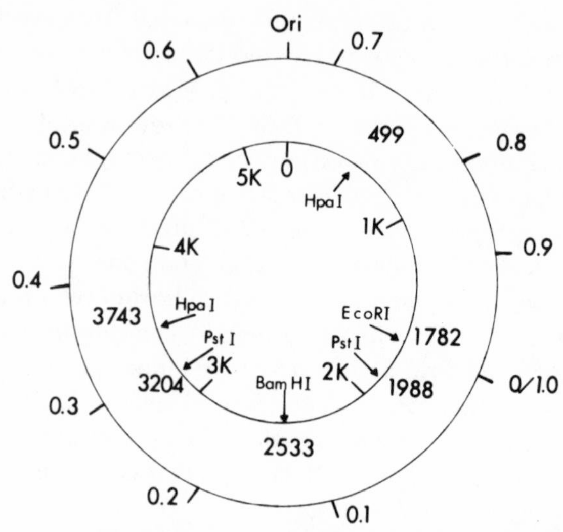

Fig. 12.1 Map of the SV40 genome. *Ori* is the origin of replication. On the outer circle are the map units, which begin at the EcoRI restriction site (map unit 0/1.0, nucleotide residue 1782). The nucleotide numbers of some other restriction sites are also shown (inner circles). The T antigen coding gene extends from nucleotide residue 5236 counterclockwise to nucleotide residue 2587. It codes for both the large and small T antigens. The promoter and enhancer for the T antigen coding gene straddle *ori*.

quently, Tjian, Fey, and Graessmann (1978) microinjected directly into the nuclei of quiescent mouse cells a purified preparation of T antigen itself, again inducing cellular DNA synthesis. These experiments convincingly demonstrated that the SV40 T antigen, alone, contained all the necessary information to initiate a new cell cycle in quiescent animal cells. This was further confirmed by the finding that a monoclonal antibody against T antigen, microinjected into the nuclei of quiescent cells, inhibits SV40-induced but not serum-stimulated DNA synthesis (Floros et al. 1981).

Galanti et al. (1981), using the manual microinjection technique and a deletion mutant of SV40 that makes a regular large T but no small t, showed that small t is not required for the stimulation of cell DNA synthesis. Furthermore, the integrity of the large T is not necessary. Some of these results are summarized in Fig. 9.2 (Soprano et al. 1983). Notice for instance that dl mutant 1001 is capable of stimulating cellular DNA synthesis. This mutant codes for a large T antigen that is only 32,000 M.W. or only ⅓ of the wild-type large T. Indeed, from the DNA sequence, one can deduce that the T coded by dl 1001 stops at amino acid residue 272 (large T has a total of 708 amino acids). Yet, this truncated T is fully capable of inducing DNA synthesis in quiescent animal cells.

While we are looking at Fig. 9.2, we should point out a few observations that are of interest, namely: (1) quite a few mutants can stimulate cell DNA synthesis although incapable of viral DNA replication; (2) the absence of a poly (A) signal or of a termination codon does not seem to drastically affect the expression of T antigen, at least in microinjected cells; (3) mutants 1001, 1055, and 1139 stimulate cell DNA synthesis but cannot reactivate silent rRNA genes in hybrid cells. Which brings us to the next topic.

Growth in Size

We have already mentioned in chapter 9 that growth in size and cell DNA replication can be dissociated. The SV40 genome offers a striking example of how these two processes can be separated. As before, for convenience we shall take RNA, especially rRNA, synthesis and/or accumulation as an indicator of growth in size. SV40 infection has been known to stimulate cellular RNA synthesis (May, May, and Bordé 1976), but its effect on rDNA genes can be best illustrated in those human–rodent hybrid cell lines we have already discussed in chapter 7. For instance, 55-54 cells are hybrids between human HT-1080 sarcoma cells and mouse peritoneal macrophages. They contain all human chromosomes and 18 mouse chromosomes, including 3 chromosomes carrying mouse rRNA genes (Croce 1976). Despite the presence of mouse rRNA genes, these hybrid cells express only human rRNA; the mouse rRNA genes are suppressed (Croce et al. 1977). Infection of 55-54 cells with SV40 causes the expression of mouse rRNA (Soprano et al. 1979). This agrees with the ability of SV40 to induce a few rounds of mitoses, since cell division usually requires growth in size.

However, quite a few deletion mutants of SV40, while capable of stimulating cell DNA synthesis, cannot reactivate silent rRNA genes, as clearly illustrated in Fig. 9.2. Thus, the information for growth in size, although encoded in the SV40 A gene, is located on sequences different from those sufficient for cell DNA replication. Indeed, from the paper by Soprano et al. (1983) it seems that at least

420 amino acids (from the amino terminus of T) are required for growth in size. These findings, incidentally, point out that T antigen, like many other proteins, is a multifunctional protein, with different functions located on different domains of the polypeptide.

Finally, no one has yet reported induction of cellular DNA synthesis or growth in size by SV40 without the concomitant expression of T antigen. Notice in Fig. 9.2 that 1046, which does not express a detectable T antigen, is deprived of any biological activity. It seems that the main antigenic determinant of SV40 T antigen is located in the same general domain as the sequences necessary of the induction of cell DNA replication.

ADENOVIRUS

The reader is again referred to the book by Tooze (1980) for general information about adenoviruses, their subgroups, structure, infectivity, etc. In this chapter we are only concerned with their ability to initiate a new cell cycle in quiescent animal cells. In this respect, the adenoviruses that have been most extensively studied are human adenoviruses 2 and 5 and, to a lesser extent, human adenovirus 12.

In permissive cells—human cells such as HeLa or KB cells—human adenoviruses actually inhibit cellular DNA synthesis (Piña and Green 1969). However in quiescent populations of semipermissive mouse or hamster cells, adenovirus (2, 5, or 12) can induce stimulation of cell DNA synthesis (Laughlin and Strohl 1976a, Rossini, Weinmann, and Baserga 1979; Braithwaite, Murray, and Bellett 1981). Under appropriate conditions (confluent monolayers), adenovirus infection can stimulate cell DNA replication even in permissive human cells (Laughlin and Strohl 1976b). However, the stimulation of cell DNA synthesis is not followed by mitosis (Braithwaite, Murray, and Bellet 1981), and in hamster cells, in fact, adenovirus infection, while inducing DNA synthesis, does not cause accumulation of RNA, i.e., growth in size (Pochron et al. 1980). It seems, therefore, that the adenovirus genome contains information for the stimulation of cell DNA replication but not for growth in size of the cell, another example of the dissociation between cellular DNA synthesis and growth in size. Fig. 9.1 is a dramatic demonstration of the ability of adenovirus infection to stimulate cellular DNA synthesis but not RNA accumulation. The reader should compare Fig. 9.1

with Fig. 1.6, which is also a scattergram from a flow microfluorimeter showing the amount of DNA and RNA in each cell. In Fig. 1.6 the cells were stimulated with serum, and even cells still in G_1 (2n amount of DNA) have a high RNA content. In Fig. 9.1, where semi-permissive hamster cells were infected with adenovirus, the cells have increased their DNA content (ordinate) but the amount of cellular RNA (abscissa) has remained at low levels.

Which gene or genes in the adenovirus genome are responsible for the induction of cellular DNA synthesis? The adenovirus genome is somewhat more complicated than the SV40 (or polyoma) genome, 35,000 base pairs against, for example, the 5,243 base pairs of SV40. Like SV40 and polyoma, adenoviruses produce early proteins and late proteins and, again, the early proteins are the ones which interest us. Even so, the number of early proteins and early mRNAs is bewildering, because of different starting points, alternative pathways of splicing, etc. The reader is referred to Flint and Broker (1980) for a list of early mRNAs and early proteins coded by the adenovirus genome. Additional information can be found in the papers by Chow, Lewis, and Broker (1980) and Galos et al. (1979). The situation is further complicated by the elaborate relationships between the various early genes and their products.

Despite these reservations, the regions of the adenovirus genome involved in the induction of cellular DNA synthesis have been identified by the use of deletion and temperature-sensitive mutants of the virus and by microinjection of adenovirus DNA restriction fragments. The findings of Rossini, Jonak, and Baserga (1981) can be summarized as follows: (1) in quiescent tsAF8 cells at the permissive temperature of 34° (tsAF8 cells are a G_1-specific ts mutant; see chapter 6), regions 1A and 2 are necessary and sufficient to initiate cellular DNA synthesis; (2) at the temperature of 40.6° (restrictive for tsAF8 cells) two additional early regions are required, 1B and 5. However, previous studies have shown that early region 1A plays a regulatory role on the transcription of the other early regions (Jones and Shenk 1979). The question therefore arises (and I quote it directly from Rossini, Jonak, and Baserga 1981): "Is the 72K protein, as a product of early region 2, the only direct requirement for the induction of cell DNA synthesis, or are the translation product(s) of region 1A also directly involved beyond their requirement for the expression of early region 2?" Braithwaite et al. (1983), working with mouse and rat cells, have found no requirement for region 2, and, although not excluding other genes, believe that early region 1A is sufficient for the induction of cellular DNA synthesis.

Whether region 1A is sufficient or both 1A and 2 are required, adenovirus-induced cellular DNA synthesis is highly irregular. As already mentioned, there is no concomitant increase in cell size, there are no mitoses, and cells can even acquire amounts of DNA larger than G_2 amounts (Braithwaite et al. 1983). While there is no question that the DNA synthesized in response to adenovirus infection of semipermissive cells is of cellular nature (Rossini, Weinmann, and Baserga 1979) one wonders whether one is dealing with a normal S phase or with some other form of cell DNA replication, for instance, gene amplification.

OTHER VIRUSES

Other DNA viruses have been less studied in terms of their ability to induce transient cell DNA synthesis, although the literature on transformation is more abundant. There are different reasons for this scarcity of information. In the case of papilloma viruses, the obstacle has been the inability to grow the viruses in tissue cultures. Recently, however, this obstacle has been removed by the successful molecular cloning of some of the papilloma viruses. In the case of Epstein-Barr virus (EBV), cytomegalovirus (CMV), and herpes simplex virus (HSV), the obstacle is the size of their genomes. These viruses have genomes that are 50 times larger than the genomes of SV40 or polyoma and the identification of the specific genes responsible for the stimulation of cellular DNA synthesis is much more difficult.

Robinson and Smith (1981) have given the most careful analysis of the induction of cellular DNA synthesis by EBV. They infected human B lymphocytes with EBV and followed the incorporation of $[^3H]$-thymidine and the percentage of labeled cells up to 96 hours. The infected cells began to synthesize DNA between 36–48 hours, reaching a maximum of 30 percent labeled cells at 96 hours. The entry into S was preceded by the appearance of the EBNA antigen, a nuclear antigen that appears in cells transformed by EBV and is coded by the EBV genome.

The evidence that papilloma viruses cause transient induction of cellular DNA synthesis is softer. For human papilloma virus, it rests on the observation of Butel (1972) that there is an increase in the incorporation of [^{3}H]-thymidine in cultures of human kidney cells exposed to virus particles. The increase varied from only twofold to fivefold higher than in control cultures.

A good stimulation of cellular DNA synthesis can be induced in human and monkey cells by infection with cytomegalovirus (St. Jeor et al. 1974). The investigators were able to show that the induced DNA synthesis was of the semiconservative type and not due to DNA repair.

A modest stimulation of cellular DNA synthesis by herpes simplex virus-2 (but not herpes simplex virus-1) in isolated rat hepatocytes has been reported by Isom (1980).

Retroviruses and pBR322

The stimulation of cellular DNA synthesis by microinjected retroviral genomes will be considered in chapter 13, while we shall discuss here the stimulation of cellular DNA synthesis by plasmid pBR322, the most commonly used vector in recombinant DNA technology. This stimulation is one of the most striking examples of a laboratory artifact, and yet it is of interest because it points out how prokaryotic gene products can be recognized by regulatory components of the eukaryotic cell. pBR322 unquestionably stimulated cellular DNA synthesis (Fig. 12.2) when microinjected into Swiss 3T3 cells (Hyland et al. 1984). It did not induce mitoses nor an increase in the amount of cellular RNA. When the tetracycline-resistance gene was absent or interrupted, the ability of the microinjected plasmid to stimulate cellular DNA synthesis was abolished. There is a sharp dependence on the cell line used: Syrian hamster cells do not respond with cell DNA synthesis to the microinjection of pBR322.

The evidence in the literature clearly indicates that pBR322 can be transcribed in mammalian cells (Kaufman and Sharp 1982) and that the transcripts can be translated into functional products (Wong, Nicolau, and Hofschneider 1980). The promoters of certain viral genes are recognized in bacterial cells and direct the expression of genes under their control (Jenkins, Howett, and Rapp 1983). This interchangeability between prokaryotic and eukaryotic regulatory elements is of interest both from a theoretical and a practical point of view. As to the ability of pBR322 to stimulate cellular DNA synthesis in Swiss 3T3 cells, one is, of course, tempted to hope that

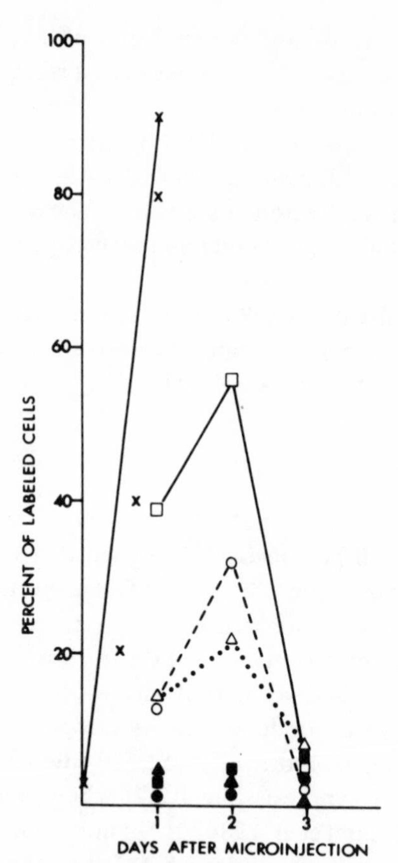

Fig. 12.2 Induction of cellular DNA synthesis by microinjection of pBR322. Quiescent Swiss 3T3 cells were microinjected with pBR322 at 0 time. Separate cultures were labeled with [^{3}H]-thymidine from 0–24 hours, from 24–48 hours, or from 48–72 hours. Open symbols: microinjected cells. Closed symbols: control cells, not microinjected, on the same coverslips as the microinjected cells. The various symbols represent three separate experiments. The crosses represent cells stimulated with 10% serum and labeled with [^{3}H]-thymidine for 24 hours. (Reprinted, with permission, from Hyland et al. 1984.)

the identification of the plasmid gene responsible for it may lead us to a better understanding of the regulation of DNA replication in eukaryotic cells. This hope sprung already when studying virus-induced cellular DNA synthesis, and it is fitting that, at this point, we should take up the discussion of the similarities (or dissimilarities) between DNA synthesis induced by growth factors or by viruses.

MECHANISMS OF VIRAL-INDUCED CELL DNA REPLICATION

Until recently it was assumed that these DNA viruses stimulated cell DNA replication by the same mechanisms that are operative in serum-stimulated cells. If the mechanism were exactly the same, this would indicate that the untransformed cell would be endowed say with a T-antigen-like protein, coded by a cellular gene, and one could envisage probable homologies between this hypothetical cellular gene and the T antigen coding gene. Unfortunately, some of the results we have already seen cast some doubts on this desirable hypothesis. The ability to stimulate cell DNA replication without concomitant growth in cell size, the absence (in some cases) of mitoses, and other irregularities of adenovirus-induced cell DNA synthesis already suggest that different pathways may be operative in serum-induced and viral-induced DNA synthesis. It is true that the kinetics of entry into S phase are about the same. For instance, with ts13 cells, the lag period between stimulation and entry into S is roughly the same, regardless of whether the stimulation has been carried out with serum or with SV40 (Floros et al. 1981). But there are other findings that clearly point to differences in the mechanisms of stimulation.

The first, and foremost, difference is the ability of SV40 adenovirus to by-pass the temperature-sensitive block in G_1-specific mutants of the cell cycle. The cell lines chosen for these studies were tsAF8 and ts13, already discussed in chapter 6. It will be remembered that when quiescent populations of tsAF8 and ts13 are stimulated with serum at the nonpermissive temperatures (39.5° – 40.6°), they do *not* enter S phase. However, infection of these cell lines with adenovirus 2 (or 5) induces cellular DNA synthesis even at the nonpermissive temperatures (Rossini, Weinmann, and Baserga 1979). Another dramatic difference between serum-stimulated and adenovirus-induced cellular DNA synthesis is their sensitivity to actinomycin D (Laughlin and Strohl 1976a,b). A concentration of 0.01 μg/ml of actinomycin D is sufficient to cause 99 percent inhibition of serum-stimulated DNA synthesis in BHK21 cells. However, 0.03 μg/ml of actinomycin D did not inhibit cellular DNA synthesis induced by infection with adenovirus 2, although virus yield was >90 percent reduced. With SV40 the situation was a little more complicated. Because tsAF8 and ts13 cells do not have receptors for SV40 and therefore cannot be infected, we microinjected them with a cloned fragment of the SV40 genome containing the T antigen coding gene. Again, cellular DNA synthesis was induced even at the nonpermissive temperatures (Floros et al. 1981). The ability of SV40 to by-pass a temperature-sensitive block in G_1 has been confirmed in G_1-specific mutants of rat 3Y1 fibroblasts (Ohno and Kimura 1984). It is therefore apparent that SV40 and adenovirus can induce cellular DNA replication in the absence of a number of cellular functions that are required by serum-stimulated cells.

SV40 can also induce cellular DNA synthesis in the presence of concentrations of butyrate that inhibit serum-stimulated DNA synthesis (Kawasaki, Diamond, and Baserga 1981). Finally, there are the senescent human diploid fibroblasts, those terminally arrested cells that were discussed in chapter 2. These senescent human diploid fibroblasts are no longer capable of entering S phase, even when repeatedly stimulated with serum. Infection with SV40 stimulates DNA synthesis in about 20 percent of the senescent cells (Ide et al. 1983). The cell fusion experiments discussed in chapter 7 also indicate that the SV40 genome can overrule G_1 blocks that are forbidding not only for growth factors but even for chemical carcinogens and RNA viruses.

When one puts together the data on cell-cycle specific ts mutants, the butyrate, the frequent dissociation between growth in size and DNA synthesis, and the effect on senescent cells, a picture emerges indicating that SV40 and adenovirus know ways of inducing cellular DNA synthesis to which cellular genes are not privy. This is not to say that the mechanisms are completely different. Indeed, most of the intracellular pathways are probably common to both growth factors and viruses, but there are differences and these should not be ignored.

Chapter 13

Genetic Basis of Cell Proliferation

The search for genes and gene products necessary for the transition of cells from a resting stage or mitosis into the S phase of the cell cycle began more than 20 years ago, with the pioneering experiments of Lieberman, Abrams, and Ove (1963) and of Baserga, Estensen, and Petersen (1965). These early experiments, crude by today's standards and based on the use of inhibitory drugs like actinomycin D or of liquid hybridization techniques, have been previously reviewed (Baserga 1976). Cell fusion experiments (see chapter 7) added more tantalizing evidence, but the first formal demonstration that RNA polymerase II-directed transcripts were necessary for the entry into S came only with the experiments already discussed in chapter 9. Briefly, a ts mutant of RNA polymerase II arrests in G_1 at the restrictive temperature; the same G_1 block can be obtained by microinjection into growing cells of α-amanitin which specifically inhibits RNA polymerase II. It is therefore reasonable to look for genes transcribed by RNA polymerase II that regulate cell cycle progression and even control cell proliferation. A survey of these genes is the object of this chapter. If the picture of genes involved in cell proliferation still has a number of gaps, it is because this area has been explored only in the past few years. But progress is so rapid that undoubtedly the picture will clarify in the near future.

The chapter will survey two categories of genes: those already known to be involved in the regulation of cell proliferation, and those whose expression is cell-cycle dependent or inducible by mitogenic stimuli. Of course, the fact that a gene is preferentially expressed in a phase of the cell cycle does not mean that it actually regulates cell cycle progression. However, since we do not know yet which genes do indeed control cell proliferation, it is expedient for the moment to survey all those genes that *may* be relevant to cell cycle progression, in the hope that at least some will turn out to be those we are searching for. In fact, one can go one step further and state that the gene all of us would like to identify and isolate is *the* gene (or genes) that triggers a quiescent cell into the cell cycle, the animal and cellular equivalent, so to speak, of the SV40 large T antigen, or of the yeast "start" genes. Mention of the SV40 large T antigen raises the question of how cell cycle genes may relate to transformation. It is for this reason that I have included oncogenes in this chapter (see below).

Another question that arises whenever one talks about genes that regulate cell proliferation is: what is their biochemical function? In some cases (for instance, DNA ligase) it is known, but in the great majority of cases all we know is that their expression is cell-cycle dependent or that their function is necessary for cell cycle progression. For the moment, one has to be contented with this ambiguity and incompleteness, but it is not difficult to predict that in the next few years the eukaryotic cell cycle will be solidly based on a series of genes, whose function is to regulate cell proliferation through well established biochemical mechanisms.

For convenience, I am dividing this survey into four sections: genes for the cell cycle of yeast, retroviral genes, cellular oncogenes, and genes for the animal cell cycle. For simplicity, I will refer to cell division cycle genes as cdc genes, a nomenclature borrowed from the original studies of Hartwell and collaborators (1971; 1976; 1978) on yeasts.

CDC GENES IN YEASTS

In chapter 6 we discussed the cell cycle mutants of yeast analyzed by Hartwell (1971; 1976) and by others. Some of these mutants were listed in Fig. 6.1, together with the presumptive position in the yeast cell cycle. Some of the genes responsible for the ts mutations, cdc genes, have been cloned and in a few cases their biochemical function has been determined. We will consider them briefly, beginning with the "start" genes and extending to other cdc genes.

As previously mentioned, in yeast, certain genes have been shown to have a role in a major cell cycle control point called "start." In *S. cerevisiae,* four genes (cdc 28, 36, 37, and 39) are necessary for start, but cdc 28 seems to be the one responsible for the major rate-limiting step of the cell cycle regulating the rate of cell division. In *S. pombe* there are two start genes, cdc 2 and cdc 10 (Beach, Durkacz, and Nurse 1982). The cdc 28 gene has been isolated and cloned by Nasmyth and Reed (1980) and was identified by its ability to complement the cdc 28 ts phenotype. The cdc 2 gene has been cloned by Beach, Durkacz, and Nurse (1982), who also showed that "*S. cerevisiae* cdc 28 can complement cdc 2 mutations of *S. pombe,* suggesting that the start events have a similar molecular basis in the two organisms." However, by Southern blot analysis, there was very little sequence homology between cdc 2 and cdc 28, a fact that Beach, Durkacz, and Nurse (1982) did not find surprising, because the two yeasts are not closely related and other genes are seemingly not homologous. I still find it surprising because I thought that genes controlling cell proliferation, being rather important, would be highly conserved (see for instance the cellular oncogenes). But exceptions may be as important as the rule. Three other genes that function at the start point of *S. cerevisiae* have been identified and cloned by Breter et al. (1983): cdc 36, cdc 37, and cdc 39. These genes and cdc 28 belong to four unlinked complementation groups. It would seem therefore that at least four genes are involved in the start control point of the cell cycle of *S. cerevisiae.* If I am allowed an extrapolation, I predict that also in mammalian cells more than one gene will be found to control the transition from a resting to a proliferating stage. Breter et al. (1983) have also calculated the intracellular steady-state abundance of the mRNA species corresponding to the four start genes. The values they give, all very low, in mRNA copies per haploid cell are: 1.5 ± 1 for cdc 36, 3.1 ± 1.5 for cdc 37, 4.6 ± 2 for cdc 39, and 7.0 ± 2 for cdc 28. This is discouraging news for those in search of start genes in mammalian cells.

Another cdc gene of *S. cerevisiae*, cdc 8 has been cloned by Kuo and Campbell (1983). The product of this gene is required throughout the period of DNA synthesis and is a protein ~ 34,000 to 40,000 daltons that binds to single-stranded DNA and stimulates DNA polymerase I activity (Arendes, Kim, and Sugino 1983). Kuo and Campbell (1983) have reported that the minimum DNA fragment capable of complementing the cdc 8 mutation is only 750 bp long, which would account for a 27,000 dalton protein. Perhaps, as with SV40 T antigen coding gene, the entire protein is not necessary for its main function.

The *S. cerevisiae* gene cdc 9 has been cloned by Barker and Johnston (1983) and is contained within a 3,300 base pair fragment of DNA. Available evidence indicates that a defective DNA ligase is responsible for the cdc 9 mutation, which places this gene among those of the DNA synthesizing machinery (see below). The cloned cdc 9 of *S. cerevisiae* also complements the cdc 17 mutation of *S. pombe*.

According to our definition of cdc genes, histone genes are also cdc genes because their expression is largely cell cycle dependent. Histone genes and their expression will be discussed at length below in the section on cdc genes of animal cells; however, they have also been studied in yeast. Hereford et al. (1981) investigated mRNA levels of H2A and H2B histone genes in the yeast strain SKQ2m as a function of cell cycle stage. They found that histone mRNA could be detected in significant quantities only in S-phase cells and concluded that histone mRNA levels are tightly and coordinately regulated throughout the cell cycle at both transcriptional and post-transcriptional levels.

Although the genes have not yet been cloned, two other mutants should be mentioned in this section. One is the cdc 2 mutant of *S. cerevisiae* (which is totally different from the cdc 2 mutant of *S. pombe* mentioned above). This mutant, at the restrictive temperature, fails to replicate approximately one-third of its nuclear genome

(Conrad and Newlon 1983), perhaps because it is defective in an initiation factor for DNA synthesis. The other mutant is the RNA polymerase I ts mutant described by Yamashita and Fukui (1980) in *Rhodosporidium toruloides.* This is an embarassing mutant, for me, because at the restrictive temperature it stops in G_1, and in chapter 9 I went to great lengths to show that the accumulation and/or synthesis of ribosomal RNA is not a prerequisite for the entry of mammalian cells into S phase. The simplest explanation for this contradiction is that the mechanisms regulating cell cycle progression are different in yeast (or *Neurospora*) and in animal cells. This is not a wild idea, since other aspects of the cell cycle are different (mating, growth factors, budding, etc.) and we already know that SV40 and adenoviruses can use different pathways from those of growth factors to induce cellular DNA synthesis in animal cells. Alternative explanations are possible but more complicated.

One brief comment should be made about the number of genes that may control cell cycle progression in yeasts. From the number of ts mutants isolated, it would seem that about 40 are certainly needed, but this is likely to be an underestimate. As more and more laboratories become involved in the isolation and identification of cdc genes, one can predict that the number will increase sharply. We will return to this problem when we discuss the cdc genes of animal cells.

The cell cycle dependency of gene transcription in *Chlamydomonas* can also be discussed in this section, for convenience. Ares and Howell (1982) have analyzed gene expression during the cell cycle of *Chlamydomonas reinhardi* both by two-dimensional gel electrophoresis and by RNA blot hybridization. Several changes in mRNA levels were found, and one mRNA coding for products identified as tubulins was increased 10 fold just before or during division.

RETROVIRAL ONCOGENES

We shall see at the end of this chapter that those genes of animal cells whose expression is cell-cycle dependent or that are known to regulate cell cycle progression may overlap with the so-called oncogenes. Furthermore, cellular oncogenes definitely play a role in cell proliferation and in transformation. Before discussing cellular oncogenes, though, it is necessary to consider retroviral genes. The remainder of the chapter therefore will be divided into three subheadings: retroviral oncogenes, cellular oncogenes, and cdc genes of animal cells, as previously defined. Although these three classes of genes must be considered separately, we will see that there is some overlap among them, and perhaps much more than simple overlap. Let us first examine the retroviral oncogenes. My discussion is based on the reviews by Duesberg (1983) and Bishop (1978; 1983) and is limited to those characteristics of retroviral oncogenes that are relevant to our understanding of cellular oncogenes and cdc genes.

Animal retroviruses are single-stranded RNA viruses that fall into two major classes: oncogenic retroviruses with transforming (onc) genes and the almost ubiquitous class without onc genes. All nontransforming retroviruses are characterized by a virion polymerase capable of RNA-directed DNA synthesis (Baltimore 1970; Temin and Mizutani 1970). In addition to the gene (*pol*) coding for the reverse transcriptase, retroviruses also contain two other genes, *gag* and *env*, coding for viral proteins (Bishop 1978). There are two classes of transforming retroviruses. The Rous sarcoma virus (RSV) RNA genome is the best example of class I and contains, besides the three essential genes *gag*, *pol*, and *env*, a transforming gene (onc gene) called v-*src*. A simplified structure of a retrovirus (no onc) and of RSV can be found in Duesberg (1983). The 5′ and 3′ regions of retroviruses have noncoding RNA sequences, the long terminal repeats (LTR), about which more later. Retroviruses do not replicate as RNA fragments. As stated by Bishop (1978): "Studies on the physiology of retrovirus replication led to the conclusions that the synthesis of virus-specific DNA was required for the initiation of infection and that DNA served as the template for synthesis of progeny viral RNA." Transformation by RSV (and other oncogenic retroviruses) also requires the copying of viral RNA into DNA, which is then integrated into the genome of the host cell in a manner analogous to transformation by DNA tumor viruses. Once the DNA copy is integrated into the cellular genome, it is transcribed (the LTR is both an efficient promoter and an enhancer; Gorman et al. 1982) and the product(s) of the transforming gene causes tumors in animals and transformation of cells in culture.

RSV with its complete RNA retroviral genome plus the transforming v-*src* gene is an exception. Other transforming retroviruses belong to class II and have deletions in their genome, partially replaced by the transforming gene. The presence of deletions makes it impossible for these retroviruses to replicate on their own. As stated by Duesberg (1983): "The onc genes of all other known retroviruses are part of physically independent but replication defective RNA genomes that are replicated by non-defective helper viruses." The v-*src* of RSV and the replication-defective retroviruses can easily be lost during passage in the animal or in cultured cells, and this is probably the reason why epidemics of transforming retroviruses have never been observed, although they are highly oncogenic when injected into animals. A partial list of the known transforming retroviruses and their onc genes is given in Table 13.1.

From the point of view of cellular proliferation, the most important thing to remember is that most viral oncogenes have cellular homologues, i.e., that both avian and mammalian DNAs contain nucleotide sequences closely related to the viral oncogenes. The only exception is the spleen-focus-forming virus of mice whose transforming gene does not seem to be a derivative of a cellular oncogene (Bishop 1983). The cellular homologues of viral oncogenes (v-onc) have been called c-onc (Bishop 1983) or proto-oncogenes (Duesberg 1983). In this book, we will follow the most common use and call them c-onc, while the term "cellular oncogenes" will comprise both c-onc and those transforming genes, detectable by DNA transfection, that are not homologous to v-onc genes (see below).

Our interest in this section deals with the role that c-onc genes may have in the control of cell proliferation, and therefore we can only mention briefly the intriguing problem of whether or not the v-oncs of retrovirus have derived from the c-onc genes. The majority of people believe that v-oncs are simply cellular genes transduced by retroviruses (for a review, see Bishop 1983), but Duesberg (1983) has provided a cogent critique of this hypothesis. It suffices to say that v-oncs and c-oncs have a high degree of homology but are not exactly alike. When compared to c-oncs, v-oncs have point mutations, deletions, and insertions, many of the changes being especially striking in the 3′ end of the genomes. c-onc genes also have intervening sequences, which are of course absent in the viral genome. The conclusions of Duesberg (1983) are worth quoting, because they are relevant to the possible function that c-onc's may have in cell proliferation: "Available structural evidence favours the hypothesis that either deletion and mutation of cellular sequences or the addition of virus-specific information, such as Δ gag, or both are needed to confer transforming function on proto-onc sequences transduced by

Table 13.1. Identified transforming retroviruses and their onc genes.

Viruses	Oncogenes[a]	Tumorigenicity	Transformation in cell culture
Avian viruses			
Rous sarcoma	*src*	Sarcoma	Fibroblasts
Fujinami sarcoma	Δ*gag-fps*	Sarcoma	Fibroblasts
Yamaguchi sarcoma	Δ*gag-yes*	Sarcoma	Fibroblasts
Rochester-2 sarcoma	Δ*gag-ros*	Sarcoma	Fibroblasts
Myelocytomatosis MC29	Δ*gag-myc*	Sarcoma, carcinoma, leukemia	Fibroblasts
Avian erythroblastosis	Δ*gag-erb*^{A}/*erb*^{B}	Leukemia	Fibroblasts
Myeloblastosis	Δ*gag-myb*-Δ*env*	Leukemia	Blood cells
Reticuloendotheliosis	*rel*	Leukemia	Blood cells
Mammalian viruses			
Moloney murine sarcoma	*mos*	Sarcoma	Fibroblasts
Harvey and Kirsten rat sarcoma	*ras*	Sarcoma, leukemia	Fibroblasts
Abelson murine leukemia	Δ*gag-abl*	Leukemia	Blood cells
FBJ murine osteosarcoma	*fos*	Sarcoma	Fibroblast
ST and GA feline sarcoma	Δ*gag-fes*	Sarcoma	Fibroblast
SM feline sarcoma	Δ*gag-fms*	Sarcoma	Fibroblast
Simian sarcoma	Δ*env-sis*	Sarcoma	Fibroblast

Adapted from Duesberg (1983).
a. Δ = deletion

viruses. In some cases, elements from several cellular genes appear necessary to generate an oncogenic virus. That is to say, qualitative changes are the norm."

CELLULAR ONCOGENES

While virologists were so cleverly elucidating the nature of transforming retroviruses, other laboratories came up with a startling discovery — that transfection of untransformed NIH 3T3 cells with DNA from chemically transformed cells caused the appearance, in the cultures, of foci of transformed cells whose DNA, in turn, was capable of transforming other untransformed NIH 3T3 cells (Shih et al. 1979). For the rest of the story, also in order to avoid an inordinate number of references, I must rely on the reviews by Cooper (1982) and by Land, Parada, and Weinberg (1983a).

Transformation by DNA Transfection

As pointed out by Cooper (1982), transfer of biologically active DNA, first demonstrated in bacteria in 1944, was subsequently extended to papovaviruses and then to cellular DNA. In a typical assay 10^6 recipient cells are exposed to 20 μg of donor DNA. Although DNA from normal cells did transform NIH 3T3 cells, it did so at a much lower efficiency than DNA from chemically transformed cells or from tumor cell lines. DNA from cells transformed by normal cell DNA or by tumor cell DNA induced transformation at higher efficiencies "in secondary transfection assays, indicating that the transformed cells contained activated transforming genes that could be efficiently transmitted by transfection" (Cooper 1982). When the donor DNA was of human origin, it was possible to demonstrate, by the presence of human DNA repetitive sequences (which are highly species specific) that transformation was mediated by human DNA. The presence of species-specific sequences also facilitated the cloning of these transforming genes, as they were called by Cooper (1982), or cellular oncogenes, as they are more commonly called.

It is of no use to list the many cell lines in which cellular oncogenes have been identified that are capable of transforming NIH 3T3 cells. So many are added so rapidly that any list would become quickly obsolete. Cellular oncogenes have been isolated from the DNA of neoplastic cells of chicken, mouse, rat, rabbit, and human origin. The cell lines used were human tumor cell lines, or cell lines transformed by chemicals or viruses (see Cooper 1982). Among the neoplastic cell lines were lines derived from carcinomas of the colon, lung, bladder, pancreas, skin, and breast; sarcomas; neuroblastomas; glioblastomas; and a variety of hematopoietic neoplasms (Land, Parada, and Weinberg 1983a).

Although the real significance of these transforming genes has not yet been agreed upon, there is no question that the elucidation of the phenomenon of NIH 3T3 transformation by cellular DNA is one of the most spectacular demonstrations of the powers of molecular biology applied to animal cells. From a rather obscure observation the investigators proceeded, step by step, until the actual transforming gene was isolated, cloned, sequenced, shown to have homology with the v-*ras* genes of two transforming retroviruses, the Harvey sarcoma virus (Ha-*ras*) and the Kirsten murine sarcoma virus (Ki-*ras*), and finally to contain in its sequence the transforming potential. Regardless of the criticisms that have been directed against the assay used, this was no mean achievement. But let us hear directly from Weinberg and his collaborators: "The study of retroviruses and the use of transfection has allowed the delineation of two groups of cellular proto-oncogenes. The two groups are, however, not separate and distinct. Instead . . . they have some members in common . . . the Ki-*ras* oncogene carried by Kirsten murine sarcoma virus is homologous to oncogenes detected by transfection in the DNA of human lung and colon carcinomas. The Ha-*ras* oncogene of Harvey murine sarcoma virus is the homolog of the well-studied oncogene of the human EJ/T24 bladder carcinoma cell line" (Land, Parada, and Weinberg 1983a). But we have seen before that the homologues of v-oncs are normally present in normal cells. How can they cause transformation? Before we attempt to answer this question, let us see where the known c-onc genes map on human chromosomes, because part of the answer lies in their position.

Mapping of Oncogenes on Human Chromosomes

Fig. 13.1, taken from Yunis (1983), shows a map of human chromosomes in terms of oncogenes, immunoglobulin genes, fragile sites (where breaks frequently occur) and consistent chromosome defects in human neoplasias. According to Yunis, the 15 oncogenes known in the cellular genome of eukaryotes map as follows: c-*myb* at band q23 of chromosome 6; c-*mos*, 8 q22 (band 22 of the long arm of chromosome 8); c-*myc*, 8 q24; c-*abl*, 9 q34; c-Ha-*ras*, chromosome 11; c-Ki-*ras*, chromosome 12; c-*fes*, 15 q24 q25; c-*src*, chromosome 20; c-*sis* chromosome 22. Other oncogenes, in addition to those shown in Fig. 13.1, have also been mapped on human chromosomes. N-*ras*, the third *ras*-related gene, cloned from a neuroblastoma cell line, maps on chromosome 1 (Ryan et al. 1983), and, as can be seen from Table 1 of the review by Land, Parada, and Weinberg (1983a), c-*fms* (McDonough feline sarcoma virus) on chromosome 5; c-*raf*

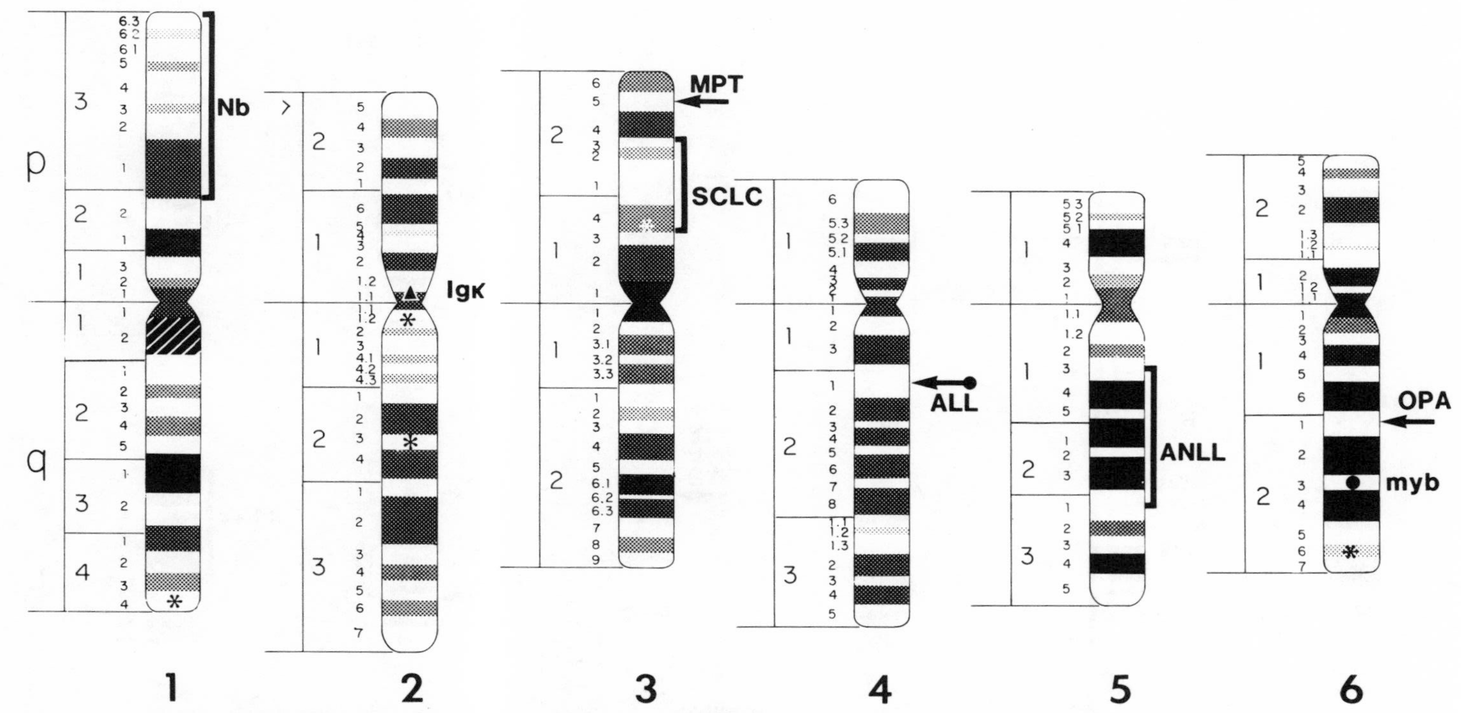

Fig. 13.1 (See the following pages for the continuation of the figure and for an explanation of the abbreviations.)

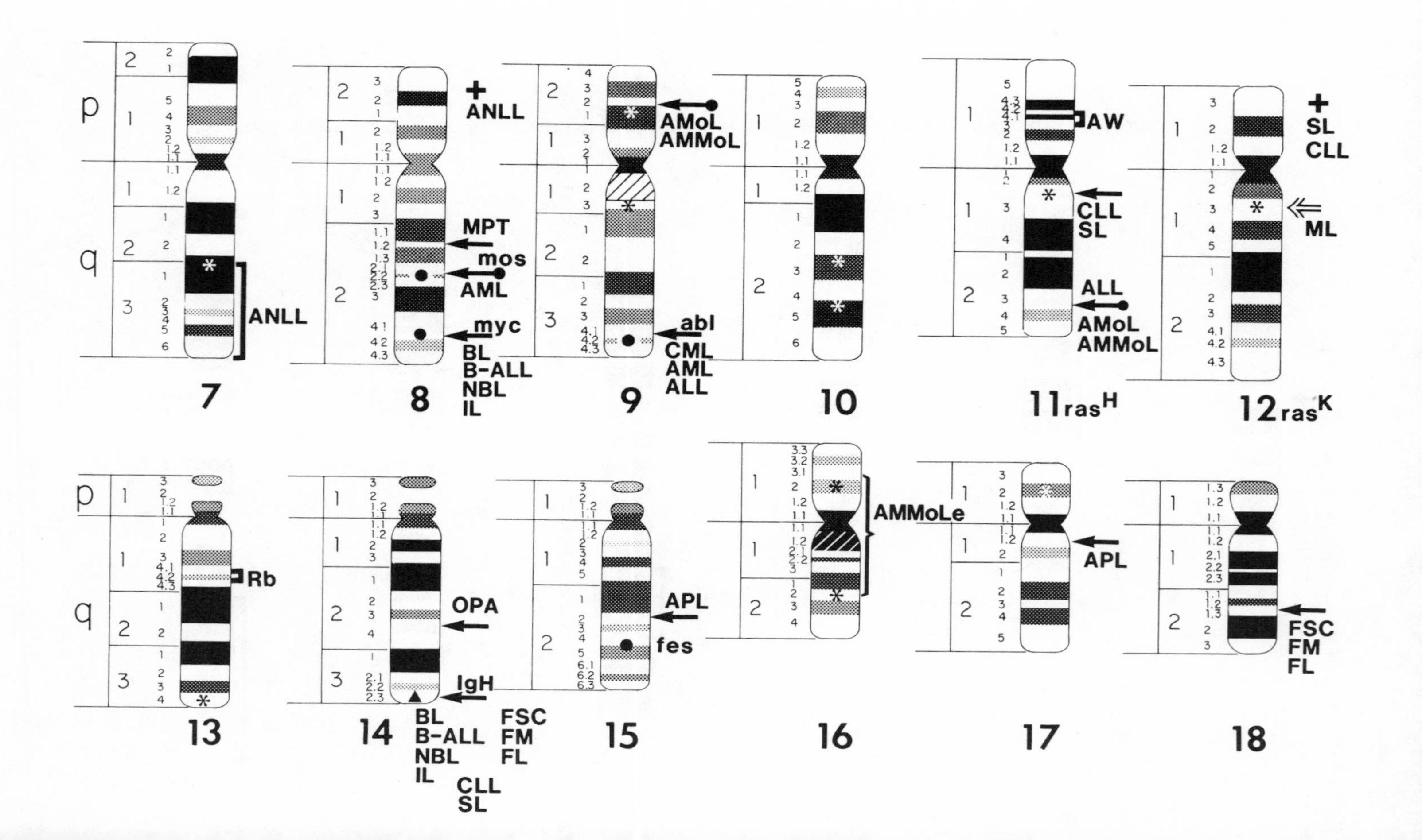
p
q
7
ANLL
8
+
ANLL
MPT
mos
AML
myc
BL
B-ALL
NBL
IL
9
AMoL
AMMoL
abl
CML
AML
ALL
10
11 ras^H
AW
CLL
SL
ALL
AMoL
AMMoL
12 ras^K
+
SL
CLL
ML
13
Rb
14
OPA
IgH
BL
B-ALL
NBL
IL
FSC
FM
FL
CLL
SL
15
APL
fes
16
AMMoLe
17
APL
18
FSC
FM
FL

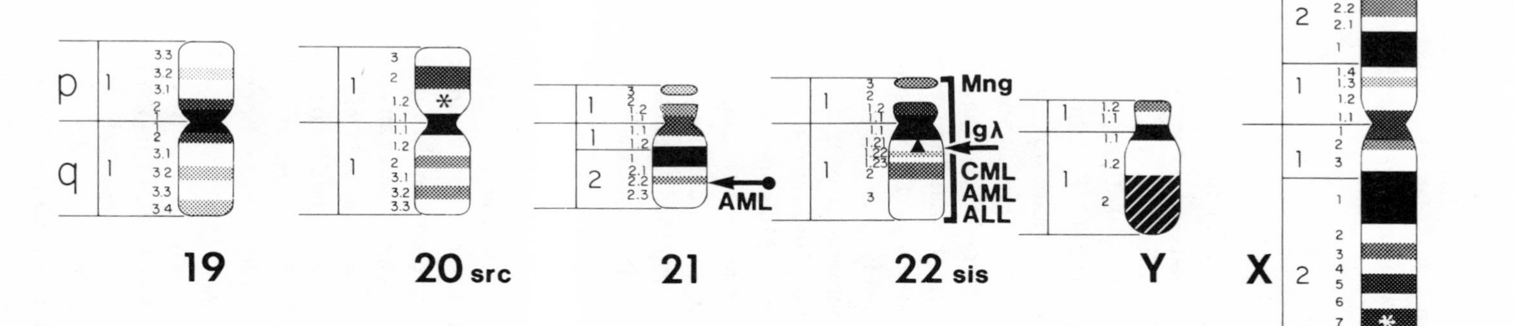

Fig. 13.1 Human chromosome map of oncogenes (large dots), fragile sites (asterisks), immunoglobulin genes (triangles), and consistent chromosome defects in human neoplasia. The karyotype represents Giemsa bands at the 400 band stage. Beginning with chromosome 1, abbreviations denote the following: Nb, neuroblastoma; Igk, kappa light chain imunoglobulin genes; MPT, mixed parotid gland tumor with t(3;8); SCLC, small cell lung cancer; ALL, acute "lymphocytic" leukemia with t(4;11); ANLL, acute nonlymphocytic leukemia; OPA, ovarian papillary adenocarcinoma with t(6;14); *mos*, Moloney sarcoma oncogene; BL, B-ALL, NBL, and IL, Burkitt's lymphoma, B cell type ALL, small noncleaved non-Burkitt's lymphoma, and immunoblastic lymphoma, respectively, with t(8;14); AMoL and AMMoL, acute monocytic and acute myelomonocytic leukemia with t(9;11); *abl*, Abelson oncogene; CML, chronic myelogenous leukemia with t(9;22); ML and broken arrows, not well defined malignant lymphoma associated with a t(12;14); AW, aniridia 'Wilms' tumor syndrome; CLL and SL, chronic lymphocytic leukemia and small lymphocytic lymphoma, respectively; ras^{H}, *ras* Harvey oncogene identified at 11p; ras^{K}, Kirsten sarcoma oncogene identified on chromosome 12; Rb, retinoblastoma; IgH, heavy-chain immunoglobulin genes; *fes*, Snyder-Theilin feline sarcoma oncogene; AMMoLe, acute myelomonocytic leukemia with increased eosinophils and inversion 16; FSC, FM, and FL, follicular small cleaved cell, follicular mixed, and follicular large cell lymphomas, respectively, with t(14;18); *src*, Rous sarcoma virus oncogene; Mng, meningioma; *sis*, Simian sarcoma oncogene; Igλ, immunoglobulin light λ chain genes. IgK and Ig are involved in Burkitt's lymphoma variant with t(2;8) or t(8;22), respectively. Heritable fragile sites (asterisks) are found in Xq27, 2q11, 9p21, 10q23, 10q25, 11q13, 12q13, 16p124, 16q22, 17p12, and 20p11. Constitutional fragile sites occur in 1q44, 2q23, 3p14, 6q26, 7q31, 9q13, and 13q34. (Reprinted, with permission, from Yunis 1983, p. 232; copyright 1983 by the American Association for the Advancement of Science.)

(3611 murine sarcoma virus) on 3; c-*fas* on 2; c-*ski* (avian SK7700 virus) on 1; and oncogenes *erb* A and B on chromosomes 17 and 7, respectively (Spurr et al. 1984).

Mechanism of Activation of c-onc Genes

Knowledge of the location of oncogenes allows us to discuss the mechanisms by which a proto-oncogene, a normal c-onc gene, can be activated and become a transforming gene. This discussion is mostly based on the reviews by Bishop (1983), Duesberg (1983), Land, Parada, and Weinberg (1983a), and Leder et al. (1983). The following mechanisms have been invoked to explain the transforming activation of normal c-onc genes: (1) Insertional mutagenesis, in which a c-onc is activated by the integration, near it, of viral DNA, especially a viral promoter. For instance, in B cell lymphomas induced in chickens by avian leukosis viruses (ALVs), the viral DNA is integrated near the c-*myc* oncogene and presumably causes its overexpression. Even more impressive was the demonstration by Blair et al. (1981) that the LTR of Moloney sarcoma virus, fused to c-*mos* (the cellular homolog of the transforming gene of Moloney sarcoma virus), conferred to the otherwise harmless c-*mos* the ability to transform cells in culture. (2) Amplification of the c-onc gene. A typical example is the c-Ki-*ras* gene, which is amplified 30 to 60 fold in cells of the mouse adrenocortical tumor Y1 (Schwab et al. 1983). The amounts of c-Ki-*ras*-specific mRNA and the p21 protein encoded by the amplified gene were correspondingly elevated, suggesting that "amplification and enhanced expression of cellular oncogenes may contribute to the genesis and/or maintenance of at least some naturally occurring tumors." Land, Parada, and Weinberg (1983a) list other cases of gene amplification: c-*myc* in the human promyelocytic leukemia cell line HL-60, c-Ki-*ras* in a human colon carcinoma cell line, N-*myc* in human neuroblastomas, and c-*abl* in a human chronic myelogenous leukemia cell line. (3) Translocation of the c-onc gene. This possibility is discussed in detail in the review by Leder et al. (1983), who have studied the translocation of c-*myc* in Burkitt lymphoma. Fig. 13.2, shows a diagram of the human chromosomes involved in the specific translocations of Burkitt lymphoma. The cellular oncogene c-*myc* is translocated to a region encoding one of the immunoglobulin genes (Della-Favera et al. 1982). The level of c-*myc* transcripts is usually, but not always, elevated in Burkitt cell lines. To explain some of the contradictory results,

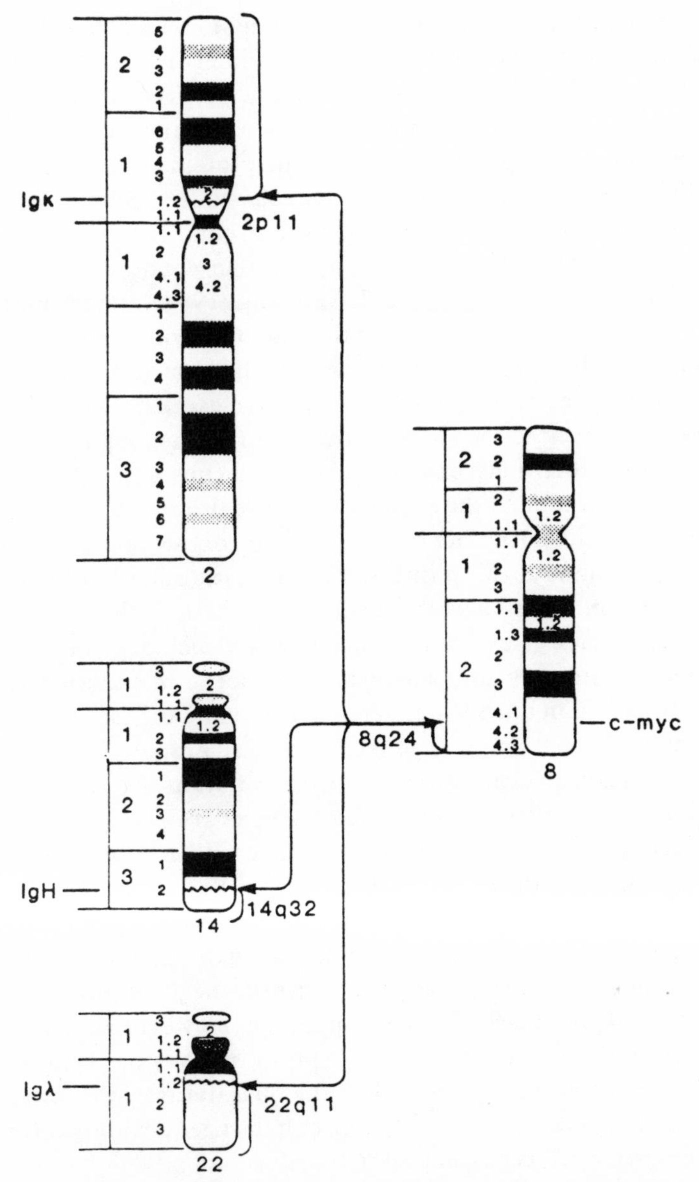

Fig. 13.2 Diagrammatic representation of the human chromosomes involved in the translocation of Burkitt lymphoma. The positions of the Igk, IgH, and Igλ chains and the c-*myc* genes are indicated. The arrows point to breakpoints at which chromosomes 2, 14, or 22 reciprocally exchange chromosomal segments with chromosome 8. (Reprinted, with permission, from Leder et al. 1983, p. 767; copyright 1983 by the American Association for the Advancement of Science.)

Leader et al. (1983) have proposed a model in which "an elevated level of c-*myc* protein produced from the deregulated, translocated allelle in a Burkitt cell could bring about the repression of the normal c-*myc* allele." (4) In the three instances mentioned above, an increased level of expression of the c-onc protein is either demonstrated or postulated. But a fourth mechanism of oncogene activation depends, not on the amount of protein, but on alterations in the amino acid sequence of the oncogene protein. Again, let us follow the review of Land, Parada, and Weinberg (1983a): "This mechanism is most well documented in the case of the oncogene proteins encoded by the *ras* genes. In the case of the human bladder carcinoma oncogene of the T24/EJ cell line, it is clear that a simple point mutation converted the Ha-*ras* proto-oncogene into a potent oncogene. This G to T transversion caused the glycine, normally present as the 12th residue of the encoded 21,000-dalton protein, to be replaced by a valine. Another activated version of this gene encodes an aspartate residue at this position." The mutations within codon 12 of the human K-*ras* locus that lead to activation of the K-*ras* gene have been reviewed by Santos et al. (1984) and include changes from glycine to cysteine, valine, and arginine. Other point mutations have been described in the c-Ki-*ras* gene. In a human lung carcinoma the mutation affected amino acid 61 of the p21 protein (Yuasa et al. 1983), while in an ovarian carcinoma the mutation affected the electrophoretic mobility of the c-Ki-*ras* product (Feig et al. 1984). A variant of the mutational changes is the mechanism by which c-*fos* is activated. According to Miller, Curran, and Verma (1984), this involves both the addition of a transcriptional enhancer sequence and a change at the 3′ end of the gene. In summary, there is good evidence that c-onc genes can acquire transforming potential by either mutations or increased levels of expression. However, on the basis of their findings with c-*fos*, Miller, Curran, and Verma (1984) have proposed "an alternative model to the quantitative or qualitative models of neoplastic transformation, that oncogene expression in an inappropriate cell type can lead to transformation."

It should be noted that other cellular oncogenes are *not* related to v-oncs. Certain oncogenes, like B*lym* (bursal lymphomas), *mam* (mammary carcinomas), and *neu* (rat neuro- and glioblastomas), are known only from transfection. B*lym* has been isolated both from chicken B cell lymphoma DNA and from Burkitt's lymphoma cell lines. Both genes are small (<1kb), have similar nucleic acid sequences, readily transform NIH 3T3 cells, and show significant homology to the amino-terminal region of the transferrin family of proteins (Diamond et al. 1983). Despite the fact that B*lym*-1, has been isolated from Burkitt's lymphoma cells, it has nothing in common with c-*myc*, and even maps on a different chromosome, specifically, on the short arm of human chromosome I (Morton et al. 1984). Another interesting group of oncogenes have been described by Lane, Sainten, and Cooper (1982). These authors induced transformation of NIH 3T3 cells with DNAs of B and T lymphocyte neoplasms of both human and mouse origin. Although the transforming genes were not cloned, restriction endonuclease analysis of the susceptibilities of the transforming activities showed that the same (or closely related) transforming genes were activated in different neoplasms in the same stage of normal cell differentiation. However, different transforming genes were activated in neoplasms from different stages of normal B and T lymphocyte differentiation.

Two Oncogenes Are Required for Full Transformation

One of the strongest criticisms of the relationship between transformation by DNA transfection and neoplasia has been that NIH 3T3 cells are not really normal cells. They are an established cell line, grossly aneuploid, and they are known to form transformed foci at the slightest provocation. Primary cultures are not transformed by transfection with individual cellular oncogenes. However, this criticism should be silenced by more recent experiments. Thus, rat embryo fibroblasts (REF) did not form foci when transfected with EJc-Ha-*ras* DNA, i.e., with the *ras* oncogene. However, they formed numerous foci when they were co-transfected with the *ras* and the *myc* oncogene (Land, Parada, and Weinberg 1983b). The *ras* oncogene could be substituted for by the gene of the polyoma virus, that codes for the middle T antigen. Conversely, the *myc* oncogene could be replaced by the gene coding for the large T antigen of polyoma virus. Similar results were obtained by Ruley (1983), who worked with primary baby rat kidney cells (BRK). These cells could not be induced to form foci by transfection with the Ha-*ras-1* gene or with gene EIA of adenovirus 2, but co-transfection with these two genes caused the appearance of numerous foci. Again, the *ras* oncogene could be substituted for by the gene for polyoma middle T antigen, or by gene E1B of adenovirus 2. Immortalization of cells, i.e., establishment of a cell line, has the same effect as Ad2-EIA (Newbold and Overell 1983).

These results have allowed Weinberg and associates to formulate the hypothesis that at least two oncogenes are necessary for the transformation of cells in culture. One class of transforming genes includes *myc*, polyoma large T, and Ad2-EIA. These genes have been implicated in the immortalization of cells in culture, i.e., the establishment of cell lines and are known to specify proteins that bind to nuclear structures. The second class of transforming genes includes Ha-*ras*, N-*ras*, and polyoma middle T. Their gene products localize to the plasma membrane. This scheme does fit neatly in the evidence for tumor progression that we have already discussed in chapter 3. It also fits with the results one can obtain with SV40. The SV40 T antigen coding gene (see chapter 12) is sufficient to cause by itself full transformation of cells in culture. In this respect, it is interesting to note that while most of the SV40 T antigen is localized in the nucleus, a small but significant amount can be found in the membrane (Soule, Lanford, and Butel 1980). Deletion mutants lacking the SV40 sequences that stimulate rRNA synthesis do not transform but can immortalize rat embryo cells (Colby and Shenk 1982). One is tempted to speculate that, in the case of SV40, the domain responsible for cellular DNA synthesis is capable of immortalizing cells but that full transformation requires also the domain that causes growth in size. Genes inducing cellular DNA synthesis and growth in size must also be present in normal cells, but here again we return to the overexpression of cellular oncogenes as a cause of transformation. In the case of SV40, it would be its promoter that would cause excessive production of the transforming protein. If this were true, one could envisage an experiment in which the SV40 T antigen coding gene, without its own promoter, could be fused to a weak promoter and thereby loose its transforming potential.

In the case of EBV, which has been discussed in the previous chapter and which shows a striking association with African Burkitt's lymphoma, that virus too is known to immortalize B lymphocytes. One can explain the association between EBV and African Burkitt's lymphoma by hypothesizing that infection by EBV predisposes B lymphocytes to full transformation by B*lym*-1 or c-*myc*, or both (Diamond et al. 1983).

Further support to the contention that oncogenes play an important role in neoplasia comes from the experiments of Sukumar et al.

(1983), who reported reproducible single-point mutations at codon 12 of the H-*ras-1* locus in DNA of mammary carcinomas induced in rats by nitroso-methylurea. Even more striking is the report by Santos et al. (1984) of an activated Ki-*ras* gene in a squamous cell carcinoma of the lung of a 66-year-old man, while the point mutation was absent in the Ki-*ras* gene from the adjacent normal tissues or the blood lymphocytes of the same patient.

While a role of cellular oncogenes in human cancer becomes more and more likely, one should not forget the possibility that some human cancers may be caused by the deletion of "suppressor" genes, i.e., genes that normally suppress the expression of the transforming genes. A good example of this theory can be found in the thoughtful review of human retinoblastoma by Murphree and Benedict (1984).

Functions of Cellular Oncogenes

Several v-onc's have tyrosine-specific protein kinase activity (see reviews by Bishop 1983; Land, Parada, and Weinberg 1983a). These include v-*src*, v-*yes*, v-*fes*, v-*abl* and v-*ros*. A tyrosine-specific protein kinase activity has been tentatively demonstrated in c-*src* and c-*fes*. This activity is particularly interesting since certain mitogenic growth factors are also known to cause phosphorylation of tyrosine (see chapter 10). To quote from Bishop (1983): "The enzymatic activity and subcellular location of pp60$^{c\text{-}src}$ therefore suggest that the protein may participate in the control of cell growth and division." Bishop himself readily admits that the evidence for this supposition is only circumstantial. At any rate, other c-onc gene (or v-onc gene) products do not display detectable protein kinase activity, and the products of *myc*, *myb*, and *fos* localize in the nucleus. It will be remembered from chapter 10 that there is no absolute correlation between tyrosine phosphorylation and the mitogenic activity of PDGF and EGF. We may add here that polyoma middle T can lose some of its phosphorylated tyrosine sites without losing its transforming ability (Oostra et al. 1983).

The expression of some c-onc genes is developmentally regulated. For instance, in the mouse, the expression of c-*fms* increases progressively in the embryonal tissues reaching a maximum at day 16 of prenatal development (Muller et al. 1983). The expression of c-Ha-*ras* remains constant throughout prenatal development, while the expression of c-Ki-*ras* declines rapidly after day 16. Expression of these c-onc genes also varies from tissue to tissue, for instance, c-*fms* is expressed in the placenta but not in the fetus. Similarly, the expression of the c-*src* gene varies with the developmental stage of the mouse embryo and is highest in neural tissues (Cotton and Brugge 1983).

Probably more interesting for the readers of this book is any possible relationship of c-onc genes to cell cycle progression. In this respect, the most seminal finding was almost simultaneously reported by two groups of investigators—that the sequence of the v-*sis* gene product, the $p28^{sis}$ protein, was the same as the sequence of PDGF (Doolittle et al. 1983; Waterfield et al. 1983). Although this has already been mentioned in chapter 10, it is worth reiterating here, because (with other findings discussed below) it brings together growth factors and cell division cycle genes, a concept that is the main focus of this book.

The homology between the v-*sis* product and PDGF offers an attractive explanation for the transforming activity of simian sarcoma virus, but—even more important—indicates that certain cellular oncogenes or cdc genes may regulate cell cycle progression by coding for growth factors or (as we shall see later) for receptors for growth factors. It should be noted that PDGF-like substances have also been detected in cells transformed by SV40 and in human osteosarcoma cells (see the discussion in the paper by Waterfield et al. 1983), while mRNAs homologous to v-*sis* have been detected in human fibrosarcoma cell lines but not in normal fibroblasts (Eva et al. 1982). Indeed, cells transformed by simian sarcoma virus contain a growth factor activity that is identical to PDGF in immunoassay, in mitogenic dose response, and in specific mitogenic activity (Deuel et al. 1983). Interestingly, the v-*sis* product is found in cellular lysates as well as in the medium, although PDGF itself is a secreted glycoprotein. This raises the question whether certain growth factors or growth factor receptors may act directly on gene expression and cell DNA replication in the cell nucleus.

The expression of certain c-onc genes is cell-cycle dependent. The first report was by Goyette et al. (1983), who observed an increased expression of the c-*ras* gene in regenerating rat liver after partial hepatectomy. The increase, detected by dot blot hybridization, was significant but modest, and occurred at the time of activation of DNA synthesis. Much more spectacular was the increase in c-*myc* expression reported by Kelly et al. (1983) in mitogen-stimulated

Table 13.2. Stimulation of cellular DNA synthesis by microinjected v-*ras* onc gene.[a]

Treatment	Percentage of cells labeled	
	Microinjected	Nonmicroinjected
v-*ras* oncogene	44	1.3
Δ v-*ras* onc	3	1.4
SV40	46	1.4

a. Quiescent Swiss 3T3 cells were microinjected and then labeled with [^{3}H]-thymidine for 48 hours. In each case, the total number of cells and the number of labeled cells in the microinjected circle were first counted. Then 10 fields were selected at random on the same coverslip and the number of labeled cells per field was determined. In each field, the total number of cells counted was equal to the total number of cells in the microinjected circle. Δ v-*ras* is similar to v-*ras*, except that the sequences coding for the transforming protein are almost completely deleted.

lymphocytes and 3T3 cells stimulated by PDGF. Differences of 20 to 40 fold between G_0 and stimulated cells were reported.

The v-*ras* oncogene, when microinjected into quiescent cells, causes a modest but reproducible stimulation of DNA synthesis as seen in Table 13.2, where the results obtained with quiescent 3T3 cells are given, but similar results were obtained with other cell lines. It should be remembered, however, that under the conditions used, we introduced into cell nuclei from 1,000 to 2,000 copies of a gene, thus creating the prerequisites for overexpression.

We will return to the significance of oncogenes in cell cycle regulation after discussing the cdc genes of animal cells.

CDC GENES IN ANIMAL CELLS

Growth factors and genes (or gene products) that regulate cell proliferation are the most important aspects of cell reproduction. The rest (discussed in parts I and II of this book) only constitutes the framework, painfully built during the past 30 years by many investigators, within which the regulation of the cell cycle will be elucidated through the identification of the relevant growth factors and genes. We are now considering some of these genes, the cdc genes of animal cells as defined in the introduction of this chapter, i.e., genes that regulate cell cycle progression or whose expression is cell-cycle dependent.

The existence of ts mutants of the mammalian cell cycle (see chapter 6) by itself suggests genes that regulate cell cycle progression. However, at the moment of writing, none of these genes has been isolated and cloned. In a couple of instances we know what they are, but we do not have yet the actual genes in our test tubes. One ts mutant, tsAF8, as already related, is a mutant of RNA polymerase II, i.e., it is defective in either the synthesis, the assembly, or the stability of RNA polymerase II. Another mutant, tsA159, is defective in a 30,000-dalton polypeptide that is required for normal topoisomerase II activity (Colwill and Sheinin 1983). Somewhat more information is available on genes whose expression is cell-cycle dependent.

Known Genes Whose Expression Is Cell Cycle Dependent

The expression of histone genes, specifically of the major histone species (H1, H2A, H2B, H3, and H4), is definitely cell-cycle dependent (Stein et al. 1982; Plumb, Stein, and Stein 1983). Although a low level of expression, especially of H1, can also be detected in G_1 cells, there is no question that histone mRNA levels are much higher in S phase cells (see also DeLisle et al. 1983). The cell cycle dependence of histone gene expression is clearly illustrated in tsAF8 cells (Hirschhorn et al. 1984a). By manipulating the time of temperature shift-up, it could be shown that the expression of H2A, H2B, H3, and H4 genes was proportional to the fraction of cells capable of entering S phase.

The expression of the cellular thymidine kinase gene is also cell-cycle dependent (Fig. 13.3). Less clear is the behavior of the dihydrofolate reductase (DHFR) gene. At first glance, it seems as if expression of the DHFR gene is cell-cycle regulated, the level of DHFR mRNA in growing cells being 10 times higher than in nongrowing cells (Kaufman and Sharp 1983). Several reports have also indicated that DHFR transcripts are not detectable in G_0 cells, but are clearly detectable in stimulated cells (see for instance LaBella, Brown, and Basilico 1983). However, a closer analysis of the available data indicates that the levels of DHFR mRNA are serum-regulated rather than cell cycle (or growth) related. Thus, DHFR mRNA levels are the

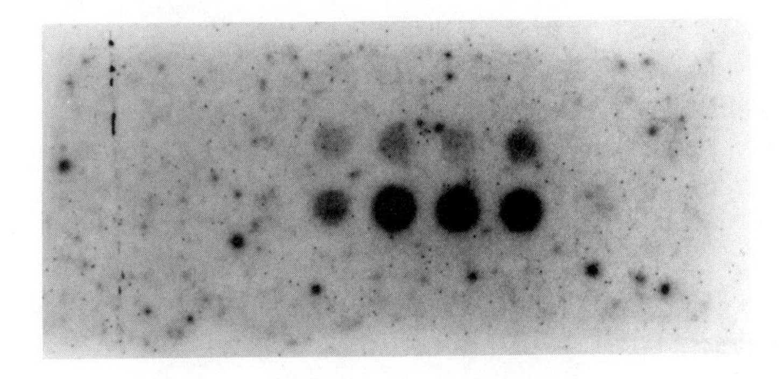

Fig. 13.3 Expression of the thymidine kinase (TK) gene during the cell cycle of tsAF8 cells. Cells were made quiescent and then stimulated with 10 percent serum at either the nonpermissive (upper row) or the permissive (lower row) temperature. RNA was isolated from cells at the indicated times (in hours) after stimulation and blotted on nitrocellulose filters. The filters were hybridized to a nick-translated DNA fragment from a cellular tk gene, p3.2. (Courtesy of John Lewis,Cold Spring Harbor Laboratory.)

same in exponentially growing cells and in nongrowing cells arrested in G_1 by starvation for isoleucine and glutamine rather than by serum (Collins et al. 1983). This raises the question (already raised by Cochran, Reffel, and Stiles 1983) of the relationship of genes inducible by growth factors to cell proliferation. My opinion is that DHFR is a cdc gene, and this opinion is based on the definition of G_0 as caused by the deficiency of growth factors, not of nutrients (see chapter 2). Returning to other known genes whose expression is cell-cycle dependent, we have already seen the marked increase in the expression of c-*myc* that occurs in mouse lympho-

Table 13.3. Known genes whose expression is cell-cycle dependent.

Gene	References
Core histones	Stein et al. (1982); Hirschhorn et al. (1984a)
Thymidine kinase	See Fig. 13.3
Dihydrofolate reductase (?)	Kaufman and Sharp (1983)
c-*ras*	Goyette et al. (1983)
c-*myc*	Kelly et al. (1983); Campisi et al. (1984)
Calmodulin	Chafouleas et al. (1984)
Actin	Campisi et al. (1984)

cytes and 3T3 cells that are stimulated to proliferate (Kelly et al. 1983). Interestingly, while c-*myc* expression is increased in proliferating 3T3 cells, in the two cell lines transformed by chemical carcinogenes, c-*myc* expression remains high, whether the cells are growing or quiescent (Campisi et al. 1984). Another gene whose expression is cell-cycle dependent is calmodulin, whose mRNA levels reach a maximum in late G_1/early S phase (Chafouleas et al. 1984). Table 13.3 summarizes the known genes in which cell-cycle dependent expression has been demonstrated.

Genes of Unknown Function Whose Expression Is Cell Cycle Dependent

Apart from the genes already mentioned above, a number of genes have been identified whose expression is cell-cycle dependent. Two such genes, for instance, have been reported by Lee et al. (1983) in K12 cells. These authors investigated the RNA levels in synchronized K12 cells of two cDNA clones, p3C5 and p4A3. Both of these genes were preferentially expressed in G_0 or early G_1, although they were still expressed (but at much lower levels) in late G_1 and S phase. Both of these genes, however, were also inducible by high temperature or glucose starvation, suggesting that they may be related to "stress" proteins. Cochran, Reffel, and Stiles (1983) screened a cDNA library for gene sequences regulated by PDGF in BALB/c 3T3 cells. Out of 8,000 clones screened, 14 independent PDGF-inducible sequences were found. One of the sequences, represented by cDNA clone JE, was represented by 100 mRNA copies per quiescent cell and increased to 3,000 copies/cell after treatment with PDGF. With another cDNA clone, KC, the corresponding numbers were 70 and 700 mRNA copies/cell. The inducibility of these two sequences was markedly enhanced by cycloheximide. It is not known yet whether these sequences are important in cell cycle progression. As pointed out by the authors themselves, "There are cellular responses to PDGF other than mitogenesis," including regulation of phospholipase activity, lipid synthesis, pinocytosis, and chemotaxis.

A similar cDNA clone was reported by Linzer and Nathans (1983). These authors screened about 3,500 clones from a cDNA library (also from BALB/c 3T3 cells) and found that 0.5 percent contained inserts corresponding to mRNAs present at higher levels in serum stimulated than in quiescent cells. One clone (28H6) hybridized to a

1-kilobase RNA species that was present at barely detectable levels in resting cells but was increased 20 fold after serum stimulation, reaching a maximum level at the G_1/S boundary. 28H6 RNA was also increased in cells stimulated by SV40 or PDGF, and the cDNA was shown to have an open reading frame coding for a protein with significant homology to mammalian prolactins.

My own laboratory has been heavily engaged in searching for similar genes, but we elected to use two G_1-specific ts mutants of the cell cycle, tsAF8 and ts13, with which the reader must be familiar since chapter 6. These two cell lines were selected in the hope that these ts mutants could be advantageously used to sort out, among the genes preferentially expressed in G_1, those that may play a major role in cell cycle progression. Using the same approach of Cochran, Reffel, and Stiles (1983) and of Linzer and Nathans (1983), we have isolated several cDNA clones containing sequences of genes whose expression, in terms of RNA levels, is increased in G_1 (Hirschhorn et al. 1984b). Fig. 13.4 shows a dot-blot in which five of these cDNA clones, nick-translated, were hybridized to cytoplasmic RNA from G_0 cells and from cells at different intervals after serum stimulation. One can see that the levels of the RNA corresponding to the inserted sequences of these clones are preferentially expressed in G_1. Some increase 6 hours after serum stimulation, and then decrease again at 16–24 hours, which, for ts 13 cells, is late G_1-S phase. Two of them (2A10 and 2A9) increase between 6 and 16 hours. mRNA corresponding to the sequences of an oncogene, c-Ki-*ras*, is present at equal levels in resting and growing cells. This is at variance with the findings of Campisi et al. (1984), who reported a cell cycle dependence of expression of the c-Ki-*ras* gene in A31 cells. The discrepancy points out how different cells regulate gene expression in different ways. Southern blots of these cDNA clones preferentially expressed in G_1 revealed that they fall into three categories: single copy or low abundance copies genes, genes belonging to multigenic families, and genes with highly repetitive sequences. From now on, therefore, we should modify our statement that the necessity of a functional RNA polymerase II for the transition of cells from G_0 (or G_1) to S (see chapter 9) demonstrates the requirement for unique copy gene transcription. It demonstrates the requirement for gene transcribed by RNA polymerase II, some of which may be unique while others may belong to multigene families or even highly repetitive sequences.

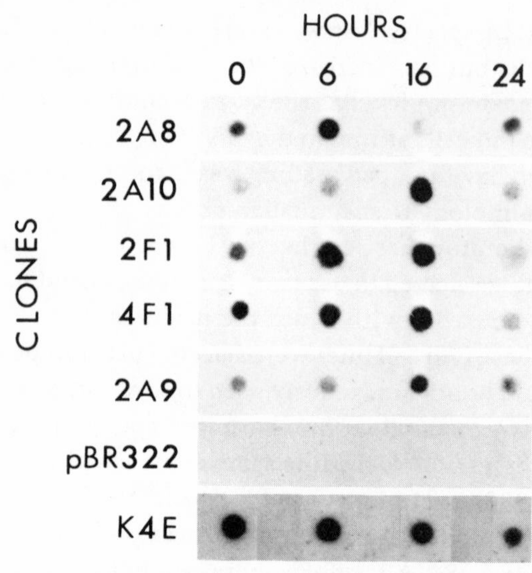

Fig. 13.4 Expression of cell-cycle specific cDNA clones after serum stimulation. Cytoplasmic poly (A+) mRNA isolated from ts13 cells (see chapter 6) at different times after serum stimulation was dotted onto nitrocellulose filters. The cDNA inserts from selected cell-cycle specific clones and control plasmids were nick-translated. After hybridization, the filters were autoradiographed. K4E (a cloned Kirsten *ras* gene) and pBR322 are included as controls. (Reprinted, with permission, from Hirschhorn et al. 1984b.)

We tested the expression of these genes in ts13 cells at the nonpermissive (Fig. 13.5) temperature, with varying results. Thus, the levels of RNA corresponding to the inserted sequences of cDNA clones 2A9 and 2F1 are increased in G_1-ts13 cells, stimulated at the nonpermissive temperatures even more than at the permissive temperature. The sequences of 2A8, instead, while elevated in G_1 cells at the permissive temperature, were not expressed in ts13 cells stimulated at the restrictive temperature. In tsAF8, a mutant of RNA polymerase II, stimulation at the nonpermissive temperature causes a rapid decrease in RNA levels corresponding to the inserted sequences of all cDNA clones, except for clone 7B5, which is *not* preferentially expressed in G_1. This indicates that these RNA preferentially expressed in G_1 have a very short half-life.

A different approach to the same problem has been taken by Schutzbank et al. (1982), who identified, in a cDNA library, 11 cDNA clones whose inserted sequences were detected in increased

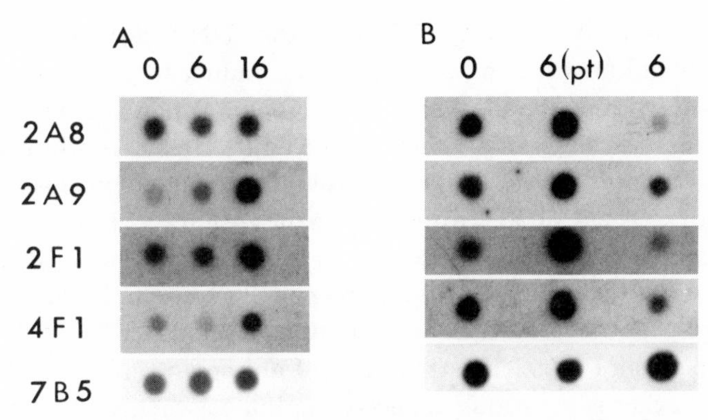

Fig. 13.5 Expression of cell-cycle specific cDNA clones after serum stimulation at the nonpermissive temperature. Same methodology as in Fig. 13.4 (A) RNA isolated from ts13 cells at 0, 6, and 16 hours after serum stimulation at the nonpermissive temperature. (B) RNA isolated from tsAF8 cells at 0 and 16 hours after serum stimulation at the permissive (34°) and 6 hours at nonpermissive temperature. The last row is a cDNA clone that is not preferentially expressed in growing cells. (Reprinted, with permission, from Hirschhorn et al. 1984b.)

amounts in 3T3 cells stimulated by SV40 infection compared to quiescent 3T3 cells. These RNA species, all cellularly coded were also increased in SV40-transformed 3T3 cells. Similar cDNA clones were described by Scott, Westphal, and Rigby (1983) in cells transformed by SV40, chemical carcinogens, or retroviruses. Since SV40 induces cell DNA synthesis by pathways that are somewhat different from those used by growth factors (see chapter 12), one would predict that the cDNA clones identified by Schutzbank et al. (1982) and by Scott, Westphal, and Rigby (1983) will be a subset of the clones that are preferentially expressed in G_1.

Other Transcripts

I will briefly enumerate in this section the data available on the appearance of new mRNA species in animal cells exposed to growth stimuli. Using BALB/c 3T3 cells, Hendrickson and Scher (1983) reported that PDGF stimulated the rapid and selective accumulation of several species of mRNAs, which were identified by cell-free translation. These mRNAs accumulated also after the addition of EGF but not after the addition of insulin or platelet-poor plasma.

The relationship of these mRNAs to the cDNA clone mentioned above is unknown. A new RNA species induced by EGF was also reported by Foster et al. (1982). This RNA was polyadenylated, its levels increased about 10 fold within 6 hours after EGF stimulation and was related to the class of mouse retrovirus or transposon-like elements termed VL30. To quote from the authors: "The physiological significance of VL30 genes is not known. They are a dispersed class of moderately repetitive sequence elements (100 copies per haploid genome) which are 5.2 Kb long, contain 0.5 Kb-LTRs and encode a major 30 S RNA transcript of unknown function." There are also other suggestions of an association between VL30 gene expression and proliferative capacity of cells, and it will be remembered that some of the serum-inducible genes described in our laboratory displayed repetitive sequences.

New mRNA species have been described in mitogen-stimulated T lymphocytes by Freeman et al. (1983), with some RNAs appearing early and disappearing late after stimulation, just as some of the RNA sequences found in ts13 and tsAF8 cells.

SOME SPECULATIONS

Having discussed the available data on the genetic basis of cell proliferation, I would like at this point to make a few impermanent (and perhaps impertinent) conclusions.

We have seen in chapters 9 and 10 data suggesting that the progression of cells through the cell cycle may depend, at least in part, on the successive actions of different growth factors. Some evidence also indicated that receptors for different growth factors do become available at different times after G_0 cells are stimulated to proliferate. The best illustration is the one given by the T lymphocytes, which we have already partially described, but that merits some further comments and reiteration.

When purified human T lymphocytes are exposed to phytohemagglutinin, they increase in size but do not enter S phase. They will enter S phase, though, when T lymphocytes exposed to PHA are treated with interleukin-2 or with macrophages (Maizel et al. 1981). In the absence of previous exposure to PHA neither macrophages nor interleukin-2 can stimulate DNA synthesis in T lymphocytes (Maizel et al. 1981; Koretzky et al. 1983). Receptors for interleukin-2, not detectable in unstimulated T lymphocytes, become readily detectable after stimulation with PHA (Koretzky et al. 1983; Neckers and Cossman 1983). The findings of Neckers and Cossman (1983) merit further discussion. These authors found that the appearance of interleukin-2 receptors was necessary for the induction of receptors for transferrin, an obligatory component for the growth of cells in serum-free media. Antibodies to interleukin-2 receptors inhibited the entry of stimulated lymphocytes into S phase but only if the cells were exposed to the antibody before the appearance of the transferrin receptors. Exposure of cells to antibodies against transferrin receptors stopped cells in the S phase (Trowbridge and Lopez 1982). It suggests a model in which the progression of T lymphocytes through the cell cycle is regulated by the successive appearance of surface receptors for different growth factors (Baserga 1984). A similar finding has been reported in 3T3 cells where Clemmons, Van Wyk, and Pledger (1980) found that exposure to PDGF, which by itself did not cause cell DNA synthesis, resulted in an increase in the number of binding sites for somatomedin C which acted as a progression factor, and induced DNA synthesis in PDGF primed cells. Similarly, in both 3T3 and 10T½ cells, a transient pre-exposure to PDGF causes a 10-fold increase in the sensitivity to the mitogenic effects of EGF (Wharton et al. 1983). My first tentative conclusion is therefore that growth factors, and growth factor receptors, play a role not only in cell proliferation in general but in the orderly progression of cells through the cell cycle.

On the other hand, it is also apparent that cell cycle progression, especially from G_1 through S and mitosis, also requires other cellular components. One just has to think about the DNA synthesizing machinery and all the various enzymes that have been discussed in chapter 11 whose function is necessary for the replication of cellular DNA. Besides the enzymes necessary for DNA synthesis, it is conceivable that other internal proteins may be necessary for cell cycle progression. A good candidate, for instance, is the p53 protein. In the already reported experiments of Mercer et al. (1982) microinjection of an antibody against the p53 protein inhibited serum-stimulated DNA synthesis in 3T3 cells. My second tentative conclusion is therefore that some internal proteins, besides growth factors and receptors for growth factors, are also necessary for orderly cell cycle progression. If these two conclusions are correct then one should predict that the cdc genes, and perhaps even some of the oncogenes, will turn out to fall into three major categories, namely (1) genes coding for growth factors, (2) genes coding for receptors for growth factors, and (3) genes coding for internal proteins, such as enzymes for the DNA synthesizing machinery, regulatory proteins (p53?), enzymes like ornithine decarboxylase, which may be necessary for cellular growth in size (Kontula et al. 1984), and other gene products whose function still has to be guessed.

The information available on genes related to cell proliferation, discussed in this chapter, is very scarce but even from this limited data some tantalizing findings have emerged which support these conclusions.

In the first place there are the striking findings of Doolittle et al. (1983) and Waterfield et al. (1983) that PDGF and the transforming protein of v-*sis* have identical sequences. It seems, therefore, that one of the cellular oncogenes, probably a cdc gene, is nothing else but a gene coding for a growth factor, and an important one at that. Diamond et al. (1983) have reported that another cellular oncogene, the B*lym*, has significant homology with the protein family of the transferrins, another growth factor. There is some homology between gastrin, a hormone that stimulates cell proliferation both in vivo and in vitro, and the middle T of polyoma virus (Baldwin 1982). But even more dramatic is the report by Lorincz and Reed (1984) that there is a 21–25 percent homology between cdc 28, the start gene of *S. cerevisiae*, and several oncogene sequences, including v-*mos*, *src*, *fes*, *fps*, *yes*, and v-*raf*. The gene described by Linzer and Nathans (1983) as inducible by serum or PDGF in quiescent 3T3 cells was found to have significant homology with prolactin, which can be considered as belonging to the family of growth factors. And so it seems that the first prediction—that some of the cellular oncogenes and cdc genes will turn out to code for growth factors—is well on its way.

The necessity for receptors for growth factors has already been mentioned above. At the moment the available evidence for overlapping between oncogenes, cdc genes, and growth factors' receptors is modest but nonetheless very exciting. Thus Downward et al. (1984) have shown that the sequences of the v-*erb*B gene product and the receptor for epidermal growth factor are quite similar. Also, one of the SV40 induced cellular genes isolated from transformed mouse cells encodes a Qa/Tla class I major histocompatibility complex antigen (Brickell et al. 1983). This gene is expressed at elevated levels in several lines of transformed fibroblasts and, even more interesting to me, in PHA-stimulated lymphocytes, although its RNA is not detectable in G_0 lymphocytes. This dependency on stimulation is in agreement with a previous observation that the expression of certain HLA antigens in human cells is cell-cycle dependent, reaching a maximum of expression in the G_2 stage of the cell cycle (Sarkar et al. 1980).

As to internal proteins it is known that the mRNA levels of p53, c-*myc*, and thymidine kinase are cell-cycle dependent, and we have seen that one of the cdc genes of yeast, cdc 9, is a structural gene for DNA ligase (Barker and Johnston 1983). The relationship between the receptor for EGF and a topoisomerase would constitute a direct link between cell surface and internal proteins.

Let us then make the prediction that as more and more genes involved in the regulation of cell proliferation become known, they will fall into the three categories mentioned above. Since everyone who writes about cell proliferation has at some point or another to make comments about cancer, in which there is an obvious alteration in cell growth, I will claim that privilege and make a very simple and perhaps naive speculation, namely, that any alteration in the structure or the expression of cdc genes may result in altered cell growth. Altered growth regulation is not cancer, but it could be the first step toward it. An alteration in a second cdc gene, as illustrated in the two-oncogenes model, could lead to full transformation. Subsequent mutations could make the transformed cell less and less responsive to growth regulation, eventually leading to the cancer cell that has been known to pathologists for more than 100 years, and that is a grossly abnormal cell.

Regardless of whether these predictions will turn out to be correct or not, let me conclude this book with a very brief summary. I hope that the reader has been able to assimilate the information I have selected to illustrate how mammalian cells proliferate, and how they progress through the cell cycle. The first two parts, Biology and Cell Biology, provide a framework upon which all future experimentation can be carried out. Of more than simple historical importance, they also give us an idea of the behavior of cell populations and of the most important biological aspects of cell division. We have seen that morphologically cell division can be reduced to three points, namely, (1) growth in size of the cell, (2) S phase, and (3) mitosis. To these three points correspond biochemically a doubling of all cellular components, cell DNA replication, and the biochemistry of the mitotic spindle. All other subdivisions of the cell cycle — G_0, several phases of G_1, and the many models that have been proposed, transition probability, competence and progression, etc., have been useful and are still useful as a conceptual framework. But from a scientific point of view they have lost much of their value, for if one accepts the definition that the goal of science is to predict, these models and these subdivisions have very little predictive value in terms of the biochemical and molecular biological mechanisms that control cell cycle progression. To identify these mechanisms the road that is open is the road of molecular biology. Clearly the identification and isolation of cdc genes and their relationship to growth factors and receptors for growth factors constitute a major task in front of us in the next few years. Together with the identification of these genes and their structure will go eventually the ability of assigning, to each of these genes, a specific biochemical function.

References/Index

References

Aarden, L. A. 1979. Revised nomenclature for antigen-nonspecific T cell proliferation and helper factors. *J. Immunol.* 123:2928–2929.

Abelson, H. T., H. N. Antoniades, and C. D. Scher. 1979. Uncoupling of RNA and DNA synthesis after plasma stimulation of G_0-arrested BALB/c 3T3 cells. *Biochim. Biophys. Acta* 561:269–275.

Abraham, J., and G. Rovera. 1980. The effect of tumor-promoting phorbol diesters on terminal differentiation of cells in culture. *Molec. Cell. Biochem.* 31:165–175.

Adelman, R. C., G. Stein, G. S. Roth, and D. Englander. 1972. Age dependent regulation of mammalian DNA synthesis and cell proliferation in vivo. *Mech. Age. Dev.* 1:49–59.

Adelstein, S. J., C. P. Lyman, and R. C. O'Brien. 1964. Variations in the incorporation of thymidine into the DNA of some rodent species. *Comp. Biochem. Physiol.* 12:223–231.

Ajiro, K., A. Zweidler, T. Borun, and C. M. Croce. 1978. Species-specific suppression of histone H1 and H2b production in human/mouse hybrids. *Proc. Natl. Acad. Sci.* 75:5599–5603.

Al-Bader, A. A., A. Orengo, and P. N. Rao. 1978. G_2 phase-specific proteins of HeLa cells. *Proc. Natl. Acad. Sci.* 75:6064–6068.

Alberghina, L., and E. Sturani. 1981. Control of growth and of the nuclear division cycle in Neurospora crassa. *Microbiol. Rev.* 45:99–122.

Allfrey, V. G., J. Karn, E. M. Johnson, and G. Vidali. 1974. Relationships between nuclear protein phosphorylation and gene activation in the cell cycle of synchronized HeLa S-3 cells. In *Control of Proliferation in Animal Cells,* eds. B. Clarkson and R. Baserga. Cold Spring Harbor Laboratory, pp. 681–700.

Antman, K. H., and D. M. Livingston. 1980. Intracellular neutralization of SV40 tumor antigens following microinjection of specific antibodies. *Cell* 19:627–635.

Arendes, J., K. C. Kim, and A. Sugino. 1983. Yeast 2-μm plasmid DNA replication in vitro: purification of the cdc 8 gene product by complementation assay. *Proc. Natl. Acad. Sci.* 80:673–677.

Ares, M., and S. H. Howell. 1982. Cell cycle stage-specific accumulation of mRNAs encoding tubulin and other polypeptides in Chlamydomonas. *Proc. Natl. Acad. Sci.* 79:5577–5581.

Ashihara, T., and R. Baserga. 1979. Cell synchronization. *Meth. in Enzymology* 58:248–262.

Ashihara, T., S. D. Chang, and R. Baserga. 1978. Constancy of the shift-up point in two temperature sensitive mammalian cell lines that arrest in G_1. *J. Cell Physiol.* 96:15–22.

Ashihara, T., F. Traganos, R. Baserga, and Z. Darzynkiewicz. 1978. A comparison of cell cycle related changes in post mitotic and quiescent AF8 cells as measured by cytofluorimetry after acridine orange staining *Cancer Res.* 38:2514–2518.

Assoian, R. K., A. Komoriya, C. A. Meyers, D. M. Miller, and M. B. Sporn. 1983. Transforming growth factor in human platelets. *J. Biol. Chem.* 258:7155–7160.

Assoian, R. K., C. A. Frolik, A. B. Roberts, D. M. Miller, and M. B. Sporn. 1984. Transforming growth factor β controls receptor levels for epidermal growth factor in NRK fibroblasts. *Cell* 36:35–41.

Atmar, V. J., and G. D. Kuehn. 1981. Phosphorylation of ornithine decarboxylase by a polyamine-dependent protein Kinase. *Proc. Natl. Acad. Sci.* 78:5518–5522.

Augenlicht, L. H., and R. Baserga. 1974. Changes in the G_0 state of WI-38 fibroblasts at different times after confluence. *Exp. Cell Res.* 89:255–262.

Ayusawa, D., K. Shimizu, H. Koyama, K. Takeishi, and T. Seno. 1983. Unusual aspects of human thymidylate synthase in mouse cells introduced by DNA-mediated gene transfer. *J. Biol. Chem.* 258:48–53.

Baldwin, G. S. 1982. Gastrin and the transforming protein of polyoma virus have evolved from a common ancestor. *FEBS Letters* 137:1–5.

Balk, S. D. 1971. Calcium as a regulator of the proliferation of normal but not of transformed chicken fibroblasts in plasma-containing medium. *Proc. Natl. Acad. Sci.* 68:271–275.

Baltimore, D. 1970. RNA-dependent DNA polymerase in virions of RNA tumor viruses. *Nature* 226:1209–1211.

Barka, T. 1965. Induced cell proliferation: the effect of isoproterenol. *Exp. Cell Res.* 37:662–679.

Barker, D. G., and L. H. Johnston. 1983. Saccharomyces cerevisiae cdc 9, a structural gene for yeast DNA ligase which complements Schizo-saccharomyces pombe cdc 17. *Eur. J. Biochem.* 134:315–319.

Barrett, J. C., and P.O.P. Ts'o. 1978. Evidence for the progressive nature of neoplastic transformation in vitro. *Proc. Natl. Acad. Sci.* 75:3761–3765.

Barski, G., S. Sorieul, and F. Cornefert. 1960. Production dans des cultures in vitro de deux souches cellulaires en association, de cellules de caractére "hybride". *C. R. Hebd. Seances Acad. Sci.* 251:1825–1827.

Baserga, R. 1976. *Multiplication and Division in Mammalian Cells.* New York: Marcel Dekker.

——— 1981. Introduction to cell growth: growth in size and DNA replication. In *Tissue Growth Factors,* ed. R. Baserga. Heidelberg: Springer–Verlag, pp. 1–12.

——— 1984. Growth in size and cell DNA replication. *Exp. Cell Res.* 151:1–5.

Baserga, R., and D. Malamud. 1969. *Autoradiography.* New York: Harper and Row.

Baserga, R., and F. Wiebel. 1969. The cell cycle of mammalian cells. *Intern. Rev. Exp. Pathol.* 7:1–30.

Baserga, R., R. D. Estensen, and R. O. Petersen. 1965. Inhibition of DNA synthesis in Ehrlich ascites cells by actinomycin D. II: the presynthetic block in the cell cycle. *Proc. Natl. Acad. Sci.* 54:1141–1148.

Baserga, R., D. Thatcher, and D. Marzi. 1968. Cell proliferation in mouse kidney after a single injection of folic acid. *Lab. Invest.* 19:92–96.

Baserga, R., D. E. Waechter, K. J. Soprano, and N. Galanti. 1982. Molecular biology of cell division. *Ann. N. Y. Acad. Sci.* 397:110–120.

Basilico, C. 1977. Temperature sensitive mutations in animal cells. *Adv. Cancer Res.* 24:223–266.

——— 1978. Selective production of cell cycle specific ts mutants. *J. Cell Physiol.* 95:367–376.

Baxter, G. C., and C. P. Stanners. 1978. The effect of protein degradation on cellular growth characteristics. *J. Cell. Physiol.* 96:139–146.

Baxter, J. D., P. H. Seeburg, J. Shine, J. A. Martial, R. D. Ivarie, L. K. Johnson, J. C. Fiddes, and H. M. Goodman. 1979. Structure of growth hormone gene sequences and their expression in bacteria and in cultured cells. In *Hormones and Cell Culture,* eds. G. H. Sato and R. Ross. Cold Spring Harbor Laboratory, pp. 317–337.

Beach, D., B. Durkacz, and P. Nurse. 1982. Functionally homologous cell cycle control genes in budding and fission yeast. *Nature* 300:706–709.

Becker, A. J., E. A. McCulloch, and J. E. Till. 1963. Cytological demonstration of the clonal nature of spleen colonies derived from transplanted bone marrow cells. *Nature* 197:452–454.

Benecke, B. J., and K. H. Seifart. 1975. DNA-directed RNA polymerase from HeLa cells: isolation, characterization and cell cycle distribution of three enzymes. *Biochem. Biophys. Acta* 414:44–58.

Bertazzoni, U., M. Stefanini, G. Pedrali-Noy, E. Giulotto, F. Nuzzo, A. Falaschi, and S. Spadari. 1976. Variations of DNa polymerases-α and -β during prolonged stimulation of human lymphocytes. *Proc. Natl. Acad. Sci.* 73:785–789.

Bishop, J. M. 1978. Retroviruses. *Ann. Rev. Biochem.* 47:35–88.

——— 1983. Cellular oncogenes and retroviruses. *Ann. Rev. Biochem.* 52:301–354.

Blair, D. G., M. Oskarsson, T. G. Wood, W. L. McClements, P. J. Fischinger, and G. G. VandeWoude. 1981. Activation of the transforming potential of a normal cell sequence: a molecular model for oncogenesis. *Science* 212:941–943.

Block, P., I. Seiter, and W. Oehlert. 1963. Autoradiographic studies of the initial cellular response to injury. *Exp. Cell Res.* 30:311–321.

Boivin, A., R. Vendrely, and C. Vendrely. 1948. L'acide desoxyribonucleique du noyau cellulaire depositaire des caractères hereditaires, arguments d'orde analitique. *C. R. Acad. Sci.* 226:1061–1063.

Bombik, B. M., and R. Baserga. 1974. Increased RNA synthesis in nuclear monolayers of WI-38 cells stimulated to proliferate. *Proc. Natl. Acad. Sci.* 71:2038–2042.

Bombik, B. M., and M. M. Burger. 1973. cAMP and the cell cycle: inhibition of growth stimulation. *Exp. Cell Res.* 80:88–94.

Bordin, S., R. C. Page, and A. S. Narayanan. 1984. Heterogeneity of normal human diploid fibroblasts: isolation and characterization of one phenotype. *Science* 223:171–173.

Bowen-Pope, D. and R. Ross. 1982. Platelet-derived growth factor. *J. Biol. Chem.* 257:5161–5171.

Braithwaite, A. W., J. D. Murray, and A. J. D. Bellett. 1981. Alterations to controls of cellular DNA synthesis by adenovirus infection. *J. Virol.* 39:331–340.

Braithwaite, A. W., B. F. Cheetham, P. Li, C. R. Parish, L. K. Waldron-Stevens, and A. J. D. Bellett. 1983. Adenovirus-induced alterations of the cell growth cycle: a requirement for expression of E1A but not of E1B. *J. Virol.* 45:192–199.

Brent, T. P., J. A. V. Butler, and A. R. Crathorn. 1965. Variations in phosphokinase activities during the cell cycle in synchronous populations of HeLa cells. *Nature* 207:176–177.

Bresciani, F. 1968. Cell proliferation in cancer. *Eur. J. Cancer* 4:343–366.

Bresciani, F., R. Paoluzi, M. Benassi, C. Nervi, C. Casale, and E. Ziparo. 1974. Cell kinetics and growth of squamous cell carcinomas in man. *Cancer Res.* 34:2405–2415.

Breter, H. J., J. Ferguson, T. A. Peterson, and S. I. Reed. 1983. Isolation and transcriptional characterization of three genes that function at start, the controlling event of the Saccharomyces cerevisiae cell division cycle: cdc 36, cdc 37 and cdc 39. *Mol. Cell. Biol.* 3:881–891.

Brickell, P. M., D. S. Latchman, D. Murphy, K. Willison, and P. W. J. Rigby. 1983. Activation of a Qa/Tla class I major histocompatibility antigen gene is a general feature of oncogenesis in the mouse. *Nature* 306:756–760.

Bucher, N. L. R. 1963. Regeneration of mammalian liver. *Intern. Rev. Cytol.* 15:245–300.

Burger, M. M. 1970. Proteolytic enzymes initiating cell division and escape from contact inhibition of growth. *Nature* 227:170–171.

Burk, R. R. 1970. One-step growth cycle for BHK 21/13 hamster fibroblasts. *Exp. Cell Res.* 63:309–316.

Burmer, G. C., P. S. Rabinovitch, and T. H. Norwood. 1984. Evidence for differences in the mechanism of cell cycle arrest between senescent and serum-deprived human fibroblasts: heterokaryon and metabolic inhibitor studies. *J. Cell. Physiol.* 118:97–103.

Burstin, S. J., H. K. Meiss, and C. Basilico. 1974. A temperature sensitive cell cycle mutant of the BHK cell line. *J. Cell Physiol.* 84:397–408.

Butel, J. S. 1972. Studies with human papilloma virus modeled after known papovaviruses systems. *J. Nat. Cancer Inst.* 48:285–299.

Campisi, J., and E. E. Medrano. 1983. Cell cycle perturbations in normal and transformed fibroblasts caused by detachment from the substratum. *J. Cell. Physiol.* 114:53–60.

Campisi, J., E. E. Medrano, G. Morreo, and A. B. Pardee. 1982. Restriction point control of cell growth by a labile protein: evidence for increased stability in transformed cells. *Proc. Natl. Acad. Sci.* 79:436–440.

Campisi, J., H. E. Gray, A. B. Pardee, M. Dean, and G. E. Sonenshein. 1984. Cell-cycle control of c-myc but not c-ras expression is lost following chemical transformation. *Cell* 36:241–247.

Carpenter, G. 1980. Epidermal growth factor is a major growth promoting agent in human milk. *Science* 210:198–199.

——— 1981. Epidermal growth factor. In *Tissue Growth Factors,* ed. R. Baserga. Heidelberg: Springer-Verlag, pp. 89–132.

Carpenter, G., C. M. Stoscheck, and A. M. Soderquist. 1982. Epidermal growth factor. *Ann. N. Y. Acad. Sci.* 397:11–17.

Chadwick, D. E., G. G. Ignotz, R. A. Ignotz, and I. Lieberman. 1980. Inhibitors of RNA synthesis and passage of chick embryo fibroblasts through the G_1 period. *J. Cell. Physiol.* 104:61–72.

Chafouleas, J. G., W. E. Bolton, H. Hidaka, A. E. Boyd III, and A. R. Means. 1982. Calmodulin and the cell cycle: involvement in regulation of cell cycle progression. *Cell* 28:41–50.

Chafouleas, J. G., L. Lagace, W. E. Bolton, A. E. Boyd III, and A. R. Means. 1984. Changes in Calmodulin and its mRNA accompany reentry of quiescent (G_0) cells into the cell cycle. *Cell* 36:73–81.

Chambard, J. C., A. Franchi, A. LeCam, and J. Pouyssegur. 1983. Growth factor-stimulated protein phosphorylation in G_0/G_1-arrested fibroblasts. *J. Biol Chem.* 258:1706–1713.

Cheetham, B. F., D. C. Shaw, and A. J. D. Bellett. 1982. Adenovirus type 5 induces progression of quiescent rat cells into S phase without polyamine accumulation. *Molec. Cell. Biol.* 2:1295–1298.

Chen, D. J., and R. J. Wang. 1982. Studies on cell division in mammalian cells. VI, A temperature-sensitive mutant blocked in both G_1 and G_2 phases of the cell cycle. *Som. Cell Genet.* 8:653–666.

Chen, H. W. 1984. Role of cholesterol metabolism in cell growth. Fed. Proc. 43:126–130.

Chen, L. B., and J. M. Buchanan. 1975. Mitogenic activity of blood components. *Proc. Natl. Acad. Sci.* 72:131–135.

Cherington, P. V., B. L. Smith, and A. B. Pardee. 1979. Loss of epidermal growth factor requirement and malignant transformation. *Proc. Natl. Acad. Sci.* 76:3937–3941.

Choie, D. D., and G. W. Richter. 1974. Cell proliferation in mouse kidney induced by lead. *Lab. Invest.* 30:647–651.

Cholon, J. J., R. G. Knopf, and R. M. Pine. 1979. Inhibition of cellular transition from G_1-resting to G_1-prereplicative phase by aminonucleoside of puromycin. *In Vitro* 9:736–742.

Chou, J. Y., and S. E. Schlegel-Hauter. 1981. Study of liver differentiation in vitro. *J. Cell Biol.* 89:216–222.

Chow, L. T., J. B. Lewis, and T. R. Broker. 1980. RNA transcription and splicing at early and intermediate times after adenovirus-2 infection. *Cold Spring Harbor Symp.* 44:401–414.

Clark, J. L., and S. Greenspan. 1979. Similarities in ornithine decarboxylase regulation in intact and enucleated cells. *Exp. Cell Res.* 118:253–260.

Clemmons, D. R., and D. S. Shaw. 1983. Variables controlling somatomedin production by cultured human fibroblasts. *J. Cell. Physiol.* 115:137–142.

Clemmons, D. R., and J. J. Van Wyk. 1981. Somatomedin: physiological control and effects on cell proliferation. In *Tissue Growth Factors,* ed. R. Baserga. Heidelberg: Springer-Verlag, pp. 161–208.

Clemmons, D. R., J. J. Van Wyk, and J. W. Pledger. 1980. Sequential addition of platelet factor and plasma to BALB/c-3T3 fibroblast cultures stimulates somatomedin-C binding early in the cell cycle. *Proc. Natl. Acad. Sci.* 77:6644–6648.

Cochet-Meilhac, M., P. Nuret, J. C. Courvalin, and P. Chambon. 1974. Animal DNA-dependent RNA polymerases. 12, Determination of the cellular number of RNA polymerase B molecules. *Biochim. Biophys. Acta.* 353:185–192.

Cochran, B. H., A. C. Reffel, and C. D. Stiles. 1983. Molecular cloning of gene sequences regulated by platelet-derived growth factor. *Cell* 33:939–947.

Cohen, L. S., and G. P. Studzinski. 1967. Correlation between cell enlargement and nucleic acid and protein content of HeLa cells in unbalanced growth produced by inhibitors of DNA synthesis. *J. Cell. Physiol.* 69:331–340.

Cohen, S. 1962. Isolation of a mouse submaxillary gland protein accelerating incisor eruption and eyelid opening in the new-born animal. *J. Biol. Chem.* 237:1555–1562.

Cohen, S. S., and H. D. Barner. 1954. Studies on unbalanced growth in Escherichia coli. *Proc. Natl. Acad. Sci.* 40:885–893.

Colby, W. W., and T. Shenk. 1982. Fragments of the simian virus 40 transforming gene facilitate transformation of rat embryo cells. *Proc. Natl. Acad. Sci.* 79:5189–5193.

Collett, M. S., and R. L. Erikson. 1978. Protein kinase activity associated with the avian sarcoma virus src gene product. *Proc. Natl. Acad. Sci.* 75:2021–2024.

Collins, J. M., M. S. Glock, and A. K. Chu 1982. Nuclease S_1 sensitive sites in parental deoxyribonucleic acid of cold and temperature-sensitive mammalian cells. *Biochemistry* 21:3414–3419.

Collins, M. L., J. R. Wu, C. L. Santiago, S. L. Hendrickson, and L. F. Johnson. 1983. Delayed processing of dihydrofolate reductase heterogeneous nuclear RNA in amino acid-starved mouse fibroblasts. *Mol. Cell. Biol.* 3:1792–1802.

Colwill, R. W., and R. Sheinin. 1983. tsA159 locus in mouse L cells may encode a novobiocin binding protein that is required for DNA topoisomerase II activity. *Proc. Natl. Acad. Sci.* 80:4644–4648.

Conrad, M. N., and C. S. Newlon. 1983. Saccharomyces cerevisiae cdc 2 mutants fail to replicate approximately one-third of their nuclear genome. *Mol. Cell. Biol.* 3:1000–1012.

Cooper, G. M. 1982. Cellular transforming genes. *Science* 217:801–806.

Cooper, J. A., D. F. Bowen-Pope, E. Raines, R. Ross, and T. Hunter. 1982. Similar effects of platelet-derived growth factor and epidermal growth factor on the phosphorylation of tyrosine in cellular proteins. *Cell* 31:263–273.

Cotton, P. C., and J. S. Brugge. 1983. Neural tissues express high levels of the cellular src gene product $pp60^{c\text{-}src}$. *Mol. Cell. Biol.* 3:1157–1162.

Crane, M. St. J., and D. B. Thomas. 1976. Cell cycle, cell shape mutant with features of the G_0 state. *Nature* 261:205–208.

Crawford, L. V., D. C. Pim, E. G. Gurney, P. Goodfellow, and J. Taylor-Papadimitriou. 1981. Detection of a common feature in several human tumor cell lines: a 53,000 dalton protein. *Proc. Natl. Acad. Sci.* 78:41–45.

Crissman, H. A., A. P. Stevenson, R. J. Kissane, and R. A. Tobey. 1979. Techniques for quantitative staining of cellular DNA for flow cytometric analysis. In *Flow Cytometry and Sorting*, eds. M. R. Melamed, P. F. Mullaney, and M. L. Mendelsohn. New York: Wiley, pp. 243–261.

Croce, C. M. 1976. Loss of mouse chromosomes in somatic cell hybrids between HT-1080 human fibrosarcoma cells and mouse peritoneal macrophages. *Proc. Natl. Acad. Sci.* 73:3248–3252.

Croce, C. M., A. Talavera, C. Basilico, and O. J. Miller. 1977. Suppression of production of mouse 28S ribosomal RNA in mouse-human hybrids segregating mouse chromosomes. *Proc. Natl. Acad. Sci.* 74:694–697.

Cunningham, D. D. 1981. Proteases as growth factors. In *Tissue Growth Factors*, ed. R. Baserga. Heidelberg: Springer-Verlag, pp. 229–248.

Cuppage, F. E., and A. Tate. 1967. Repair of the nephron following injury with mercuric chloride. *Am. J. Path.* 51:405–429.

Dalla-Favera, R., M. Bregni, J. Erikson, D. Patterson, R. C. Gallo, and C. M. Croce. 1982. Human c-myc oncgene is located on the region of chromosome 8 that is translocated in Burkitt lymphoma cells. *Proc. Natl. Acad. Sci.* 79:7824–7827.

Daniel, C. W. 1977. Cell longevity in vivo. In *Handbook of the Biology of Aging*, eds. C. E. Finch and L. Hayflick. New York: Van Nostrand, pp. 159–186.

D'Anna, J. A., R. A. Tobey, and L. R. Gurley. 1980. Concentration-dependent effects of sodium butyrate in Chinese hamster cells: cell cycle progression, inner-histone acetylation, histone H1 dephosphorylation and induction of an H1-like protein. *Biochemistry* 19:2656–2671.

Darzynkiewicz, Z., F. Traganos, and M. R. Melamed. 1980. New cell cycle compartments identified by multiparameter flow cytometry. *Cytometry.* 1:98–108.

Darzynkiewicz, Z., F. Traganos, T. Sharpless, and M. R. Melamed. 1976. Lymphocyte stimulation: a rapid multiparameter analysis. *Proc. Natl. Acad. Sci.* 73:2881–2884.

——— 1977. Cell cycle-related changes in nuclear chromatin of stimulated lymphocytes as measured by flow cytometry. *Cancer Res.* 37:4635–4640.

Darzynkiewicz, Z., D. P. Evenson, L. Staiano-Coico, T. K. Sharpless, and M. L. Melamed. 1979. Correlation between cell cycle duration and RNA content. *J. Cell. Physiol.* 100:425–438.

Darzynkiewicz, Z., T. Sharpless, L. Staiano-Coico, and M. R. Melamed. 1980. Subcompartments of the G_1 phase of the cell cycle detected by flow cytometry. *Proc. Natl. Acad. Sci.* 77:6696–6699.

Davidson, R. L., and P. S. Gerald. 1976. Improved techniques for the induction of mammalian cell hybridization by polyethylene glycol. *Som. Cell Genet.* 2:165–176.

Davidson, R. L., K. A. O'Malley, and T. B. Wheeler. 1976. Polyethylene glycol-induced mammalian cell hybridization: effect of polyethylene glycol molecular weight and concentration. *Som. Cell Genet.* 2:271–280.

Decker, J. M., and J. J. Marchalonis. 1978. Molecular events in lymphocyte activation: role of nonhistone chromosomal proteins in regulating gene expression. *Contemp. Topics in Mol. Immunol.* 7:365–413.

DeLarco, J. E., and G. J. Todaro. 1978. Growth factors from murine sarcoma virus-transformed cells. *Proc. Natl. Acad. Sci.* 75:4001–4005.

DeLeo, A. B., G. Jay, E. Appella, G. C. Dubois, L. W. Law, and L. J. Old. 1979. Detection of a transformation related antigen in chemical induced sarcomas and other transformed cells of the mouse. *Proc. Natl. Acad. Sci.* 76:2420–2424.

DeLisle, A. J., R. A. Graves, W. F. Marzluff, and L. F. Johnson. 1983. Regulation of histone mRNA production and stability in serum-stimulated mouse 3T6 fibroblasts. *Mol. Cell. Biol.* 3:1920–1929.

Dethlefsen, L. A. 1980. In quest of the quaint quiescent cells. In *Radiation Biology in Cancer Research* eds. R. E. Meyn and H. R. Withers. New York: Raven Press, pp. 415–435.

Dethlefsen, L. A., K. D. Bauer, and R. M. Riley. 1980. Analytical cytometric approaches to heterogenous cell populations in solid tumors: a review. *Cytometry* 1:89–97.

Deuel, T. F., J. S. Huang, S. S. Huang, P. Stroobant, and M. D. Waterfield. 1983. Expression of a platelet-derived growth factor-like protein in simian sarcoma virus transformed cells. *Science* 221:1348–1350.

Diamond, A., G. M. Cooper, J. Ritz, and M. A. Lane. 1983. Identification and molecular cloning of the human Blym transforming gene activated in Burkitt's lymphomas. *Nature* 305:112–116.

Doolittle, R. F., M. W. Hunkapiller, L. E. Hood, S. G. Devare, K. C. Robbins, S. A. Aaronson, and H. N. Antoniades. 1983. Simian sarcoma virus onc gene, v-sis, is derived from the gene (or genes) encoding a platelet-derived growth factor. *Science* 221:275–277.

Douglas, S. D., P. F. Hoffman, J. Borjeson, and L. N. Chessin. 1967. Studies on human peripheral blood lymphocytes in vitro. *J. Immunol.* 98:17–30.

Downward, J., Y. Yarden, E. Mayes, G. Scrace, N. Totty, P.Stockwell, A. Ullrich, J. Schlessinger, and M. D. Waterfield. 1984. Close similarity of epidermal growth factor receptor and v-erb-B oncogene protein sequences. *Nature* 307:521–527.

Dubbs, D. R., and S. Kit. 1976. Reactivation of chick erythrocyte nuclei in heterokaryons with temperature sensitive Chinese hamster cells. *Som. Cell Genet.* 2:11–19.

Duesberg, P. H. 1983. Retroviral transforming genes in normal cells? *Nature* 304:219–226.

Dulbecco, R. 1970. Topoinhibiton and serum requirement of transformed and untransformed cells. *Nature* 227:802–806.

Dulbecco, R., L. H. Hartwell, and M. Vogt. 1965. Induction of cellular DNA synthesis by Polyoma virus. *Proc. Natl. Acad. Sci.* 53:403–410.

Durham, J. P., and R. O. Lopez-Solis. 1982. The effect of isoproterenol and cycloheximide on protein synthesis and growth in mouse parotid. *Exp. Molec. Path.* 37:235–248.

Durham, J. P., R. Baserga, and F. R. Butcher, 1974. The effect of isoproterenol and its analogs upon adenosine 3′,5′-monophosphate and guanosine 3′,5′-monophosphate levels in mouse parotid gland in vivo. *Biochem. Biophys. Acta.* 372:196–217.

Eagle, H. 1955. Nutrition needs of mammalian cells in tissue cultures. *Science* 122:501–504.

Eilen, E., R. Hand, and C. Basilico. 1980. Decreased initiation of DNA synthesis in a temperature-sensitive mutant of hamster cells. *J. Cell Physiol.* 105:259–266.

Epifanova, O. I. 1977. Mechanisms underlying the differential sensitivity of proliferating and resting cells to external factors. *Intern. Rev. Cytol.* (suppl.) 5:303–335.

Eriksson, S., and D. W. Martin, 1981. Ribonucleotide reductase in cultured mouse lymphoma cells. *J. Biol. Chem.* 256:9436–9440.

Eva, A., K. C. Robbins, P. R. Andersen, A. Srinivasan, S. R. Tronick, E. P. Reddy, N. W. Ellmore, A. T. Galen, J. A Lautenberger, T. S. Papas, E. H. Westin, F. Wong-Staal, R. C. Gallo, and S. A. Aaronson. 1982. Cellular genes analogous to retroviral onc genes are transcribed in human tumor cells. *Nature* 295:116–119.

Fabrikant, J. I. 1968. The kinetics of cellular proliferation in regenerating liver. *J. Cell. Biol.* 36:551–565.

Feig, L. A., R. C. Bast, Jr., R. C. Knapp, and G. M.Cooper. 1984. Somatic activation of ras^K gene in a human ovarian carcinoma. *Science* 223:698–701.

Fernandez-Pol, J. A. 1977. Iron: a possible cause of the G_1 arrest induced in NKR cell by picolinic acid. *Biochem. Biophys. Res. Comm.* 78:136–143.

Fischer, B., G. Schluter, C. P. Adler, and W. Sandritter. 1970. Zytophotometrische DNS-, Histon- und Nicht-Histonprotein-Bestimmungen an Zell Kernen von menschlichen Herzen. *Beitr. Path.* 141:238–260.

Flint, S. J., and T. R. Broker. 1980. Lytic infection by adenoviruses. In *DNA Tumor Viruses,* ed. J. Tooze. Cold Spring Harbor Laboratory, pp. 443–546.

Floros, J. and R. Baserga. 1980. Reactivation of G_0 nuclei by S phase cells. *Cell Biol. Intern. Rep.* 4:75–82.

Floros, J., G. Jonak, N. Galanti, and R. Baserga. 1981. Induction of cell DNA replication in G_1 specific ts mutants by microinjection of recombinant SV40 DNA. *Exp. Cell Res.* 132:215–223.

Foster, D. N., L. J. Schmidt, C. P. Hodgson, H. L. Moses, and M. J. Getz. 1982. Polyadenylated RNA complementary to a mouse retrovirus-like multigene family is rapidly and specifically induced by epidermal growth factor stimulation of quiescent cells. *Proc. Natl. Acad. Sci.* 79:7317–7321.

Fox, T. O., and A. B. Pardee. 1970. Animal cells: noncorrelation of length of G_1 phase with size after mitosis. *Science* 167:80–82.

Fraser, R. S. S., and P. Nurse. 1978. Novel cell cycle control of RNA synthesis in yeasts. *Nature* 27:726–730.

Freeman, G. J., C. Clayberger, R. DeKruyff, D. S. Rosenblum, and H. Cantor. 1983. Sequential expression of new gene programs in inducer T cell clones. *Proc. Natl. Acad. Sci.* 80:4094–4098.

Frisque, R. J., D. B. Rifkin, and W. C. Topp. 1980. Requirement for the large T and small t proteins of SV40 in the maintenance of the transformed state. *Cold Spring Harbor Symp.* 44:325–331.

Fry, R. J. M., S. Lesher, and H. I. Kohn. 1962. A method for determining mitotic time. *Exp. Cell Res.* 25:469–471.

Fry, R. J. M., S. Lesher, W. E. Kisieleski, and G. Sacher. 1963. Cell proliferation in the small intestine. In *Cell Proliferation,* eds. L. F. Lamerton and R. J. M. Fry. Oxford:Blackwell, pp. 213–233.

Fukuda, M., and A. Sibatani, 1953. Biochemical studies on the number and the composition of liver cells in postnatal growth of the rat. *J. Biochem.* 40:95–110.

Galanti, N., G. J. Jonak, K. J. Soprano, J. Floros, L. Kaczmarek, S. Weissman, V. B. Reddy, S. M. Tilghman, and R. Baserga. 1981. Characterization and biological activity of cloned simian virus 40 DNA fragments. *J. Biol. Chem.* 256:6469–6474.

Galos, R. S., J. Williams, M. H. Binger, and S. J. Flint. 1979. Location of additional early gene sequences in the adenoviral chromosome. *Cell* 17:945–956.

Gates, B. J., and M. Friedkin. 1978. Mid-G_1 marker protein(s) in 3T3 mouse fibroblast cells. *Proc. Natl. Acad. Sci.* 75:4959–4961.

Gaub, J., G. Auer, and A. Zetterberg. 1975. Quantitative cytochemical aspects of a combined Feulgen naphthol yellow S staining procedure for the simultaneous determination of nuclear and cytoplasmic proteins and DNA in mammalian cells. *Exp. Cell Res.* 92:323–332.

Gelfant, S. 1977. A new concept of tissue and tumor cell proliferation. *Cancer Res.* 37:3845–3862.

Gerson, D. F., and H. Kiefer. 1983. Intracellular pH and the cell cycle of mitogen-stimulated murine lymphocytes. *J. Cell. Physiol.* 114:132–136.

Glenn, K. C., D. H. Carney, J. W. Fenton II, and D. D. Cunningham. 1980. Thrombin active site regions required for fibroblast receptor binding and initiation of cell division. *J. Biol. Chem.* 255:6609–6616.

Gordon, S., and Z. Cohn. 1971. Macrophage-melanocyte heterokaryons. II, The activation of macrophage DNA synthesis. Studies with inhibitors of RNA synthesis. *J. Exp. Med.* 133:321–338.

Gorman, C. M., G. T. Merlino, M. C. Willingham, I. Pastan, and B. H. Howard. 1982. The Rous sarcoma virus long terminal repeat is a strong promoter when introduced into a variety of eukaryotic cells by DNA-mediated transfection. *Proc. Natl. Acad. Sci.* 79:6777–6781.

Gospodarowicz, D., and J. S. Moran. 1976. Growth factors in mammalian cell culture. *Ann. Rev. Biochem.* 45:531–538.

Goyette, M., C. J. Petropoulos, P. R. Shank, and N. Fausto. 1983. Expression of a cellular oncogene during liver regeneration. *Science* 219:510–512.

Graessmann, M., and A. Graessmann. 1976. Early simian virus 40-specific RNA contains information for tumor antigen formation and chromatin replication. *Proc. Natl. Acad. Sci.* 73:366–370.

Graham, C. F., and R. W. Morgan. 1966. Changes in the cell cycle during early amphibian development. *Develop. Biol.* 14:439–460.

Grisham, J. W. 1962. A morphologic study of deoxyribonucleic acid synthesis and cell proliferation in regenerating rat liver: autoradiography with thymidine-H^3. *Cancer Res.* 22:842–849.

Gurley, L. R., R. A. Tobey, R. A. Walters, C. E. Hildebrand, P. G. Hohmann, J. A. D'Anna, S. S. Barham, and L. L. Deaven. 1978. Histone phosphorylation and chromatin structure in synchronized mammalian cells. In *Cell Cycle Regulation,* eds. J. R. Jeter, I. L. Cameron, G. M. Padilla, and A. M. Zimmerman. New York: Academic Press, pp. 37–60.

Gurney, E. G., R. O. Harrison, and J. Fenno. 1980. Monoclonal antibodies against simian virus 40 T antigens: evidence for distinct subclasses of large T antigen and for similarities among nonviral T antigens. *J. Virol.* 34:752–763.

Guzzo, J., S. Niewiarowski, J. Musial, C. Bastl, R. A. Grossman, A. K. Rao, I. Berman, and D. Paul. 1980. Secreted platelet proteins with antiheparin and mitogenic activities in chronic renal failure. *J. Lab. Clin. Med.* 96:102–113.

Ham, R. G. 1981. Survival and growth requirements of nontransformed cells. In *Tissue Growth Factors,* ed. R. Baserga. Heidelberg: Springer-Verlag, pp. 13–88.

Harris, H. 1968. *Nucleus and Cytoplasm* Oxford: Clarendon Press.

Hartwell, L. H. 1971. Genetic control of the cell division cycle in yeast. *J. Mol. Biol.* 59:183–194.

——— 1976. Sequential function of gene products relative to DNA synthesis in the yeast cell cycle. *J. Mol. Biol.* 104:803–817.

——— 1978. Cell division from a genetic perspective. *J. Cell. Biol.* 77:627–637.

Hay, R. T., and M. L. DePamphilis. 1982. Initiation of SV40 DNA replication in vivo: location and structure of 5′ ends of DNA synthesized in the ori region. *Cell* 28:767–779.

Hayflick, L. 1977. The cellular basis for biological aging. In *Handbook of the Biology of Aging,* eds. C. E. Finch and L. Hayflick. New York: Van Nostrand, pp. 159–186.

Hayflick, L., and P. S. Moorhead. 1961. The serial cultivation of human diploid cell strains. *Exp. Cell Res.* 25:585–621.

Hazelton, B., B. Mitchell, and J. Tupper. 1979. Calcium, magnesium and growth control in the WI-38 human fibroblast cell. *J. Cell Biol.* 83:487–498.

Heby, O., and J. Jänne. 1981. Polyamine antimetabolites: biochemistry, specificity and biological effects of inhibitors of polyamine synthesis. In *Polyamines in Biology and Medicine,* eds. D. R. Morris and L. J. Marton. New York: Marcel Dekker, pp. 243–310.

Heintz, N. H., and J. L. Hamlin. 1982. An amplified chromosomal sequence that includes the gene for dihydrofolate reductase initiates replication within specific restriction fragments. *Proc. Natl. Acad. Sci.* 79:4083–4087.

Heldin, C. H., A. Wasteson, and B. Westermark. 1980. Growth of normal human glial cells in defined medium containing platelet-derived growth factor. *Proc. Natl. Acad. Sci.* 77:6611–6615.

Hendrickson, S. L., and C. D. Scher. 1983. Platelet-derived growth factor-modulated translatable mRNAs. *Mol. Cell. Biol.* 3:1478–1487.

Hereford, L. M., M. A. Osley, J. R. Ludwig II, and C. S. McLaughlin. 1981. Cell-cycle regulation of yeast histone mRNa. *Cell* 24:367–375.

Herlyn, D., and H. Koprowski. 1982. IgG2a monoclonal antibodies inhibit human tumor growth through interaction with effector cells. *Proc. Natl. Acad. Sci.* 79:4761–4765.

Hidaka, H., Y. Sasaki, T. Tanaka, T. Endo, S. Ohno, Y. Fujii, and T. Nagata. 1981. N-(6-Aminohexyl)-5-chloro-1-naphthalene sulfanamide, a calmodulin antagonist, inhibits cell proliferation. *Proc. Natl. Acad. Sci.* 78:4354–4357.

Hightower, M. J., and J. J. Lucas. 1980. Construction of viable mouse-human hybrid cells by nuclear transplantation. *J. Cell. Physiol.* 105:93–103.

Hill, B. T., and L. A. Price. 1982. An experimental biological basis for increasing the therapeutic index of clinical cancer therapy. *Ann. N. Y. Acad. Sci.* 397:72–87.

Hirschhorn, R. R., F. Marashi, R. Baserga, J. Stein, and G. Stein. 1984a. Expression of histone genes in a G_1-specific temperature sensitive mutant of the cell cycle. *Biochemistry* 23:3731–3735.

Hirschhorn, R. R., P. Aller, Z. A. Yuan, C. Gibson, and R. Baserga. 1984b. Cell cycle specific cDNA's from mammalian cells temperature-sensitive for growth. *Proc. Natl. Acad. Sci.*, in press.

Hiscott, J. B., and V. Defendi. 1980. Viral and cellular control of the SV40-transformed phenotype. *Cold Spring Harbor Symp.* 44:343–352.

Hodge, L. D., T. W. Borun, E. Robbins, and M. D. Scharff. 1969. Studies on the regulation of DNA and protein synthesis in synchronized HeLa cells. In *Biochemistry of Cell Division*, ed. R. Baserga. Springfield, Ill.: Thomas, pp. 15–37.

Hodgson, G. 1967. Synthesis of RNA and DNA at various intervals after erythropoietin injection in transfused mice. *Proc. Soc. Exp. Biol. Med.* 124:1045–1047.

Hohmann, P. 1981. Histone gene expression: hybrid cells and organisms establish complex controls. *Int. Rev. Cytol.* 71:41–93.

Holley, R. W., and J. A. Kiernan. 1968. Contact inhibition of cell division in 3T3 cells. *Proc. Natl. Acad. Sci.* 60:300–304.

Holley, R. W., R. Armour, J. H. Baldwin, and S. Greenfield. 1983. Activity of a kidney epithelial cell growth inhibitor on lung and mammary cells. *Cell Biol. Int. Rep.* 7:141–147.

Hordern, J., and J. F. Henderson. 1982. Comparison of purine and pyrimidine metabolism in G_1 and S phases of HeLa and Chinese hamster ovary cells. *Can. J. Biochem.* 60:422–433.

Hori, T., D. Ayusawa, K. Shimizu, H. Koyama, and T. Seno. 1984. Chromosome breakage induced by thymidylate stress in thymidylate synthase-negative mutants of mouse FM3A cells. *Cancer Res.* 44:703–709.

Howard, A., and S. R. Pelc. 1951. Nuclear incorporation of P^{32} as demonstrated by autoradiographs. *Exp. Cell. Res.* 2:178–187.

Hsie, A. W., and T. T. Puck. 1971. Morphological transformation of Chinese hamster cells by dibutyril adenosine cyclic 3′:5′ monophosphate and testosterone. *Proc. Natl. Acad. Sci.* 68:358–361.

Huang, J. S., S. S. Huang, B. Kennedy, and T. F. Deuel. 1982. Platelet-derived growth factor. *J. Biol. Chem.* 257:8130–8136.

Hunter, T., and B. M. Sefton. 1980. Transforming gene product of Rous sarcoma virus phosphorylates tyrosine. *Proc. Natl. Acad. Sci.* 77:1311–1315.

Hyland, J. K., R. R. Hirschhorn, C. Avignolo, W. E. Mercer, M. Ohta, N. Galanti, G. J. Jonak, and R. Baserga. 1984. Microinjected pBR322 stimulates cellular DNA synthesis in Swiss 3T3 cells. *Proc. Natl. Acad. Sci.* 81:400–404.

Hyodo, M., and K. Suzuki. 1982. A temperature-sensitive mutant isolated from mouse FM3A cells defective in DNA replication at a nonpermissive temperature. *Exp. Cell Res.* 137:31–38.

Ide, T., J. Ninomiya, and S. Ishibashi. 1984. Isolation of a G_0 specific ts mutant from a Fischer rat cell line, 3Y1. *Exp. Cell Res.* 150:60–67.

Ide, T., Y. Tsuji, S. Ishibashi, and Y. Mitsui. 1983. Reinitiation of host DNA synthesis in senescent human diploid cells by infection with simian virus 40. *Exp. Cell Res.* 143:343–349.

Ingles, C. J. 1978. Temperature-sensitive RNA polymerase II mutations in Chinese hamster ovary cells. *Proc. Natl. Acad. Sci.* 75:405–409.

Ingles, C. J., and M. Shales. 1982. DNA mediated transfer of an RNA polymerase II gene: reversion of the temperature-sensitive hamster cell cycle mutant ts AF8 by mammalian DNA *Mol. Cell. Biol.* 2:666–673.

Inman, D. R., and E. H. Cooper. 1963. Electron microscopy of human lymphocytes stimulated by phytohaemagglutinin. *J. Cell Biol.* 19:441–445.

Isom, H. C. 1980. DNA synthesis in isolated hepatocytes infected with herpesviruses. *Virology* 103:199–216.

Iversen, O. H. 1981. The chalones. In *Tissue Growth Factors*, ed. R. Baserga. Heidelberg: Springer-Verlag, pp. 491–550.

Iversen, O. H., U. Iversen, J. L. Ziegler, and A. Z. Bluming. 1974. Cell kinetics in Burkitt lymphoma. *Eur. J. Cancer* 10:155–163.

Jaehning, J. A., C. C. Stewart, and R. G. Roeder. 1975. DNA dependent RNA polymerase levels during the response of human peripheral lymphocytes to phytohemagglutinin. *Cell* 4:51–57.

Jager, W., H. Rost, and P. Tautu. 1980. *Lecture Notes in Biomathematics.* 38, *Biological Growth and Spread.* Berlin: Springer-Verlag.

Jenkins, F. J., M. K. Howett, and F. Rapp. 1983. Simian virus 40 promotes direct expression of the tetracycline gene in plasmid pACYC 184. *J. of Virol.* 45:478–481.

Johnson, L. F., C. L. Fuhrman, and L. M. Wiedemann. 1978. Regulation of dihydrofolate reductase gene expression in mouse fibroblasts during the transition from the resting to growing state. *J. Cell. Physiol.* 97:397–406.

Johnson, R. T., and P. N. Rao. 1970. Mammalian cell fusion: induction of premature chromosome condensation in interphase nuclei. *Nature* 226:717–722.

Jonak, G. J., and R. Baserga. 1979. Cytoplasmic regulation of two G_1-specific temperature-sensitive functions. *Cell* 18:117–123.

——— 1980. The cytoplasmic appearance of three functions expressed during the $G_0 \rightarrow G_1 \rightarrow S$ transition is nucleus dependent. *J. Cell. Physiol.* 105:347–354.

Jones, N., and T. Shenk. 1979. An adenovirus type 5 early gene function regulates expression of other early viral genes. *Proc. Natl. Acad. Sci.* 76:3665–3669.

Junker, S., and S. Pedersen. 1981. A universally applicable method of isolating somatic cell hybrids by two-colour flow sorting. *Biochem. Biophys. Res. Comm.* 102:977–984.

Kajiwara, K., and G. C. Mueller. 1964. Molecular events in the reproduction of animal cells. II, Fractional synthesis of deoxyribonucleic acid with 5-bromodeoxyuridine and its effect on cloning efficiency. *Biochim. Biophys. Acta* 91:486–493.

Kaufman, R. J., and P. A. Sharp. 1982. Construction of a modular dihydrofolate reductase cDNa gene: analysis of signals utilized for efficient expression. *Mol. Cell. Biol.* 2:1304–1319.

——— 1983. Growth-dependent expression of dihydrofolate reductase mRNA from modular cDNA genes. *Mol. Cell. Biol.* 3:1598–1608.

Kawasaki, S., L. Diamond, and R. Baserga. 1981. Induction of cellular deoxyribonucleic acid synthesis in butyrate-treated cells by simian virus 40 deoxyribonucleic acid. *Mol. Cell. Biol.* 1:1038–1047.

Kelly, K., B. H. Cochran, C. D. Stiles, and P. Leder. 1983. Cell-specific regulation of the c-myc gene by lymphocyte mitogens and platelet-derived growth factor. *Cell* 35:603–610.

Killander, D., and A. Zetterberg. 1965. Quantitative cytochemical studies on interphase growth. *Exp. Cell Res.* 38:272–284.

Kimler, B. F., M. H. Schneiderman, and D. B. Leeper. 1978. Induction of concentration-dependent blockade in the G_2 phase of the cell cycle by cancer chemotherapeutic agents. *Cancer Res.* 38:809–814.

King, G. L., C. R. Kahn, and C. H. Heldin. 1983. Sharing of biological effect and receptors between guinea pig insulin and platelet-derived growth factor. *Proc. Natl. Acad. Sci.* 80:1308–1312.

Kishimito, S., and I. Lieberman. 1964. Synthesis of RNA and protein required for the mitosis of mammalian cells. *Exp. Cell. Res.* 36:92–101.

Kit, S., D. R. Dubbs, P. M. Frearson, and J. L. Melnick. 1966. Enzyme induction in SV40-infected green monkey kidney cultures. *Virology* 29:69–83.

Koch, K. S., and H. L. Leffert. 1979. Increased sodium ion influx is necessary to initiate rat hepatocyte proliferation. *Cell* 18:153 – 163.

Kontula, K. K., T. K. Torkkeli, C. W. Bardin, and O. A. Janne. 1984. Androgen induction of ornithine decarboxylase mRNA in mouse kidney as studied by complementary DNA. *Proc. Natl. Acad. Sci.* 81:731 – 735.

Koretzky, G. A., R. P. Daniele, W. C. Greene, and P. C. Nowell, 1983. Evidence for an interleukin-independent pathway for human lymphocyte activation. *Proc. Natl. Acad. Sci.* 80:3444 – 3447.

Korsmeyer, S. J., P. A. Hieter, J. V. Ravetch, D. G. Poplack, T. A. Waldmann, and P. Leder. 1981. Developmental hierarchy of immunoglobulin gene rearrangements in human leukemic pre-B cells. *Proc. Natl. Acad. Sci.* 78:7096 – 7100.

Kruse, P. F., Jr., E. Miedema, and H. C. Carter. 1967. Amino acid utilizations andprotein synthesis at various proliferation rates, population densities, and protein contents of perfused animal cell and tissue cultures. *Biochemistry* 6:949 – 955.

Kruse, P. F., Jr., W. Whittle, and E. Miedema. 1969. Mitotic and nonmitotic multiple-layered perfusion cultures. *J. Cell. Biol.* 42:113 – 121.

Kuo, C., and J. L. Campbell. 1983. Cloning of Saccharomyces cerevisiae DNA replication genes: isolation of the cdc 8 gene and two genes that compensate for the cdc 8 – 1 mutation. *Mol. Cell. Biol.* 3:1730 – 1737.

LaBella, F., E. H. Brown, and C. Basilico. 1983. Changes in the levels of viral and cellular gene transcripts in the cell cycle of SV40 transformed mouse cells. *J. Cell. Physiol.* 117:62 – 68.

Land, H., L. F. Parada, and R. A. Weinberg. 1983a. Cellular oncogenes and multistep carcinogenesis. *Science* 222:771 – 778.

——— 1983b. Tumorigenic conversion of primary embryo fibroblasts requires at least two cooperating oncogenes. *Nature* 304:596 – 602.

Landy-Otsuka, F., and I. Scheffler. 1980. Enzyme induction in a temperature-sensitive cell cycle mutant of Chinese hamster fibroblasts. *J. Cell. Physiol.* 105:209 – 220.

Lane, D. P., and L. V. Crawford. 1979. T antigen is bound to a host protein in SV40 transformed cells. *Nature* 278:261 – 263.

Lane, M. A., A. Sainten, and G. M. Cooper. 1982. Stage-specific transforming genes of human and mouse B- and T-lymphocyte neoplasms. *Cell* 28:873 – 880.

Larsson, A. 1969. Ribonucleotide reductase from regenerating rat liver. *Eur. J. Biochem.* 11:113 – 121.

Lau, C. C., and A. B. Pardee. 1982. Mechanism by which caffeine potentiates lethality of nitrogen mustard. *Proc. Natl. Acad. Sci.* 79:2942 – 2946.

Laughlin, C. and W. A. Strohl. 1976a. Factors regulating cellular DNA synthesis induced by adenovirus infection. I, The effects of actinomycin D on G_1-arrested BHK21 cells abortively infected with type 12 adenovirus or stimulated by serum. *Virology* 74:30 – 43.

——— 1976b. Factors regulating cellular DNA synthesis induced by adenovirus infection. II, The effects of actinomycin D on productive virus-cell systems. *Virology* 74:44 – 56.

Learned, R. M., S. T. Smale, M. M. Haltiner, and R. Tjian. 1983. Regulation of human ribosomal RNA transcription. *Proc. Natl. Acad. Sci.* 80:3558-3562.

Leder, P., J. Battey, G. Lenoir, C. Moulding, W. Murphy, H. Potter, T. Stewart, and R. Taub. 1983. Translocations among antibody genes in human cancer. *Science* 222:765-771.

Ledinko, N. 1967. Stimulation of DNA synthesis and thymidine kinase activity in human embryonic kidney cells infected by adenovirus 2 or 12. *Cancer Res.* 27:1459-1469.

Lee, A. S., A. M. Delegeane, V. Baker, and P. C. Chow. 1983. Transcriptional regulation of two genes specifically induced by glucose starvation in a hamster mutant fibroblast cell line. *J. Biol. Chem.* 250:597-603.

Lee, G. T. Y., and D. L. Engelhardt. 1977. Protein metabolism during growth of Vero cells. *J. Cell. Physiol.* 92:293-302.

Lee, J. C. K. 1971. Effects of partial hepatectomy in rats on two transplantable hepatomas. *Amer. J. Path.* 65:347-356.

Lepoint, A., and G. Goessens. 1982. Quantitative analysis of Ehrlich tumour cell nucleoli during interphase. *Exp. Cell Res.* 137:456-458.

Levine, A. J. 1978. Approaches to mapping the temporal events in the cell cycle using conditional lethal mutants. *J. Cell. Physiol.* 95:387-392.

Ley, K. D, and R. A. Tobey, 1970. Regulation of initiation of DNA synthesis in Chinese hamster cells. *J. Cell Biol.* 47:453-459.

Lieberman, I., R. Abrams, and P. Ove. 1963. Changes in the metabolism of ribonucleic acid preceding the synthesis of deoxyribonucleic acid in mammalian cells cultured from the animal. *J. Biol. Chem.* 238:2141-2149.

Lindberg, U., and T. Persson. 1972. Isolation of mRNA from KB-cells by affinity chromatography on polyuridylic acid linked to Sepharose. *Eur. J. Biochem.* 31:246-254.

Linzer, D. I. H., and A. J. Levine. 1979. Characterization of a 54K dalton cellular SV40 tumor antigen present in SV40 transformed cells and uninfected embryonal carcinoma cells. *Cell* 17:43-52.

Linzer, D. I. H., and D. Nathans. 1983. Growth-related changes in specific mRNAs of cultured mouse cells. *Proc. Natl. Acad. Sci.* 80:4271-4275.

Linzer, D. I. H., W. Maltzman, and A. J. Levine, 1979. The SV40 A gene product is required for the production of a 54,000 MW cellular tumor antigen. *Virology* 98:308-318.

Lipsich, L. A., J. L. Lucas, and J. R. Kates. 1978. Cell cycle dependence of the reactivation of chick erthrocyte nuclei after transplantation into mouse L929 cell cytoplasts. *J. Cell. Physiol.* 97:199-208.

Liskay, R. M. 1974. A mammalian somatic "cell cycle" mutant defective in G_1. *J. Cell. Physiol.* 84:49-56.

——— 1978. Genetic analysis of a Chinese hamster cell line lacking a G_1 period. *Exp. Cell Res.* 114:69-77.

Liskay, R. M., and D. M. Prescott. 1978. Genetic analysis of the G_1 period: isolation of mutants (or variants) with a G_1 period from a Chinese hamster cell line lacking G_1. *Proc. Natl. Acad. Sci.* 75:2873-2877.

Lörincz, A. T., and S. I. Reed. 1984. Primary structure homology between the product of yeast cell division control gene CDC 28 and vertebrate oncogenes. *Nature* 307:183-185.

Lydersen, B. K., F. T. Kao, and D. Pettijohn. 1980. Expression of genes coding for nonhistone chromosomal proteins in human-Chinese hamster cell hybrids. *J. Biol. Chem.* 255:3002-3007.

Macpherson, I., and L. Montagnier. 1964. Agar suspension culture for the selective assay of cells transformed by polyoma virus. *Virology* 23:291-294.

Madsen, K., U. Friberg, P. Roos, S. Eden, and O. Isaksson. 1983. Growth hormone stimulates the proliferation of cultured chondrocytes from rabbit ear and rat rib growth cartilage. *Nature* 304:545-547.

Maizel, A. L., S. R. Mehta, S. Hauft, D. Franzini, L. B. Lachman, and R. J. Ford. 1981. Human T-lymphocyte/monocyte interaction in response to lectin: kinetics of entry into the S phase. *J. Immunol.* 127:1058-1064.

Malamud, D. 1971. Differentiation and the cell cycle. In *The Cell Cycle and Cancer*, ed. R. Baserga. New York: Marcel Dekker, pp. 132-141.

Malamud, D. E., M. Gonzalez, H. Chiu, and R. A. Malt. 1972. Inhibition of cell proliferation by azathioprine. *Cancer Res.* 32:1226-1229.

Malt, R. A., and D. A. Lemaitre. 1968. Accretion and turnover of RNA in the renoprival kidney. *Am. J. Physiol.* 214:1041-1047.

Mariani, B. D., and Schimke, R. T. 1984. Gene amplification in a single cell cycle in Chinese hamster ovary cells. *J. Biol. Chem.* 259:1901-1910.

Marquardt, H., M. W. Hunkapiller, L. E. Hood, and G. J. Todaro. 1984. Rat transforming growth factor type 1: structure and relation to epidermal growth factor. *Science* 223:1079-1082.

Martin, G. M., C. A. Sprague, T. H. Norwood, and W. R. Pendergrass. 1974. Clonal selection, attenuation and differentiation in an in vitro model of hyperplasia. *Am. J. Pathol.* 74:137-154.

Massagué, J., L. A. Blinderman, and M. P. Czech. 1982. The high affinity insulin receptor mediates growth stimulation in rat hepatoma cells. *J. Biol. Chem.* 257:13958-13963.

Matsuhisa, T., and Y. Mori. 1981. An anchorage-dependent locus in the cell cycle for the growth of 3T3 cells. *Exp. Cell Res.* 135:393-398.

May, P., E. May, and J. Bordé. 1976. Stimulation of cellular RNA synthesis in mouse kidney cell cultures infected with SV40 virus. *Exp. Cell Res.* 100:433-436.

McClain, D. A., and G. M. Edelman. 1980. Density-dependent stimulation and inhibition of cell growth by agents that disrupt microtubules. *Proc. Natl. Acad. Sci.* 77:2748-2752.

McCracken, A. 1982. A temperature-sensitive DNA synthesis mutant isolated from the Chinese hamster ovary cell line. *Som. Cell Genet.* 8:179-195.

McKeehan, W. L., and K.A. McKeehan. 1981. Extracellular regulation of fibroblast multiplication: a direct kinetic approach to analysis of role of low molecular weight nutrients and serum growth factors. *J. Supramol. Struct. Cell. Biochem.* 15:83–110.

Melamed, M. R., P. F. Mullaney, and M. L. Mendelsohn. 1979. *Flow Cytometry and Sorting.* New York: Wiley.

Mendelsohn, M. L. 1962. Autoradiographic analysis of cell proliferation in spontaneous breast cancer of C3H mouse. III, The growth fraction. *J. Nat. Cancer Inst.* 28:1015–1029.

Mendelsohn, M. L., and M. Takahashi. 1971. A critical evaluation of the fraction of labeled mitoses method as applied to the analysis of tumor and other cell cycles. In *The Cell Cycle and Cancer,* ed.R. Baserga. New York: Marcel Dekker, pp. 58–95.

Mercer, W. E., and R. A. Schlegel. 1980. Cell-cycle re-entry of quiescent mammalian nuclei following heterokaryon formation. *Exp. Cell Res.* 128:431–438.

——— 1982. Cytoplasts can transfer factor(s) that stimulate quiescent fibroblasts to enter S phase. *J. Cell. Physiol.* 110:311–314.

Mercer, W. E., C. Avignolo, and R. Baserga. 1984. The role of the p53 protein in cell proliferation as studied by the microinjection of monoclonal antibodies. *Mol. Cell. Biol.* 4:276–281.

Mercer, W. E., D. Nelson, A. B. DeLeo, L. J. Old, and R. Baserga. 1982. Microinjection of monoclonal antibody to protein p53 inhibits serum-induced DNA synthesis in 3T3 cells. *Proc. Natl. Acad. Sci.* 79:6309–6312.

Mercer, W. E., D. Nelson, J. K. Hyland, C. M. Croce, and R. Baserga. 1983. Inhibition of SV40-induced cellular DNA synthesis by microinjection of monoclonal antibodies. *Virology* 127:149–158.

Mercer, W. E., C. Avignolo, N. Galanti, K. M. Rose, J. K. Hyland, S. T. Jacob, and R. Baserga. 1984. Cellular DNA replication is independent from the synthesis or accumulation of ribosomal RNA. *Exp. Cell Res.* 150:118–130.

Metcalf, D. 1981. Hemopoietic colony stimulating factors. In *Tissue Growth Factors,* ed. R. Baserga. Heidelberg: Springer-Verlag, pp. 343–384.

Miller, A. D., T. Curran, and I. M. Verma. 1984. c-fos protein can induce cellular transformation: a novel mechanism of activation of a cellular oncogene. *Cell* 36:51–60.

Milner, J., and S. Milner. 1981. SV40-53K antigen: a possible role for 53K in normal cells. *Virology* 112:785–788.

Ming, P. M. L., H. L. Chang, and R. Baserga. 1976. Release by human chromosome 3 of the block at G_1 of the cell cycle in hybrids between ts AF8 hamster and human cells. *Proc. Natl. Acad. Sci.* 73:2052–2055.

Ming, P. M. L., B. Lange, and S. Kit. 1979. Association of human chromosome 14 with a ts defect in G_1 of Chinese hamster K12 cells. *Cell Biol. Int'l. Rep.* 3:169–178.

Mironescu, S., and K. A. O. Ellem. 1977. Secondary activities of diverse inhibitors potentiate the response of hamster embryo cultures to a mitotic stimulus. *J. Cell. Physiol.* 90:281–294.

Miska, D., and H. B. Bosmann. 1980. Existence of an upper-limit to elongation of the prereplicative period in confluent cultures of C3H/10T½ cells. *Biochem. Biophys. Res. Comm.* 93:1140–1145.

Mitchison, J. M. 1971. *The Biology of the Cell Cycle.* Cambridge: Cambridge University Press.

Molls, M., N. Zamboglou, and G. Streffer. 1983. A comparison of the cell kinetics of pre-implantation mouse embryos from two different mouse strains. *Cell Tiss. Kinet.* 16:277–283.

Momparler, R. L. 1972. Kinetic and template studies with 1-β-arabino-furanosylcytosine 5′-triphosphate and mammalian deoxyribonucleic acid polymerase. *Molec. Pharmacol.* 8:362–370.

Moorehead, P. S., and V. Defendi. 1963. Asynchrony of DNA synthesis in chromosomes of human diploid cells. *J. Cell Biol.* 16:202–209.

Mora, M., Z. Darzynkiewicz, and R. Baserga. 1980. DNA synthesis and cell division in a mammalian cell mutant temperature sensitive for the processing of ribosomal RNA. *Exp. Cell Res.* 125:241–249.

Morton, C. C., R. Taub, A. Diamond, M. A. Lane, G. M. Cooper, and P. Leder. 1984. Mapping of the human Blym-1 transforming gene activated in Burkitt lymphomas to chromosome 1. *Science* 223:173–175.

Moser, G. C., and H. K. Meiss. 1982. Nuclear fluorescence and chromatin condensation of mammalian cells during the cell cycle with special reference to the G_1 phase. In *Genetic Expression in the Cell Cycle,* eds. G. M. Padilla and K. S. McCarty, Sr. New York: Academic Press, pp. 129–147.

Moses, H. L., and R. A. Robinson. 1982. Growth factors, growth factors receptors and cell cycle control mechanisms in chemically transformed cells. *Fed. Proc.* 41:3008–3011.

Moses, H. L., J. A. Proper, M. E. Volkenant, D. J. Wells, and M. J. Getz. 1978. Mechanism of growth arrest of chemically transformed cells in culture. *Cancer Res.* 38:2807–2812.

Mueller, C., A. Graessmann, and M. Graessmann. 1978. Mapping of early SV40-specific functions by microinjection of different early viral DNA fragments. *Cell* 15:579–585.

Muller, R., D. J. Slamon, E. D. Adamson, J. M. Tremblay, D. Muller, M. J. Cline, and I. M. Verma. 1983. Transcription of c-onc genes c-ras^{ki} and c-fms during mouse development. *Mol. Cell. Biol.* 3:1062–1069.

Mummery, C. L., J. Boonstra, P. T. van der Saag, and S. W. de Laat. 1981. Modulation of functional and optimal $(Na^+ - K^+)$ ATPase activity during the cell cycle of neuroblastoma cells. *J. Cell. Physiol.* 107:1–9.

Murphree, A. L., and W. F. Benedict. 1984. Retinoblastoma: clues to human oncogenesis. *Science* 223:1028–1033.

Nadal-Ginard, B. 1978. Commitment, fusion and biochemical differentation of a myogenic cell line in the absence of DNA synthesis. *Cell* 15:855–864.

Nakano, M. M., T. Sekiguchi, and M. Yamada. 1978. A mammalian cell mutant with temperature sensitive thymidine kinase. *Som. Cell Genet.* 4:169–178.

Nakano, S., S. A. Bruce, H. Ueo, and P. O. P. Ts'o. 1982. A qualitative and quantitative assay for cells lacking postconfluence inhibition of cell division: characterization of this phenotype in carcinogen-treated Syrian hamster embryo cells in culture. *Cancer Res.* 42:3132–3137.

Nasmyth, K. A., and S. I. Reed. 1980. Isolation of genes by complementation in yeasts: molecular cloning of a cell-cycle gene. *Proc. Natl. Acad. Sci.* 77:2119–2123.

Neal, J. V., and C. S. Potten. 1981. Circadian rhythms in the epithelial cells and the pericryptal fibroblast sheath in three different sites in the murine intestinal tract. *Cell Tis. Kinet.* 14:581–587.

Neckers, L. M., and J. Cossman. 1983. Transferrin receptor induction in mitogen-stimulated human T lymphocytes is required for DNA synthesis and cell division and is regulated by interleukin-2. *Proc. Natl. Acad. Sci.* 80:3494–3498.

Newbold, R. F. and R. W. Overell. 1983. Fibroblast immortality is a prerequisite for transformation by EJc-Ha-ras oncogene. *Nature* 304:648–651.

Newport, J., and M. Kirschner. 1982. A major developmental transition in early Xenopus embryos. I, Characterization and timing of cellular changes at the midblastula stage. *Cell* 30:675–686.

Nias, A. H. W., and M. Fox. 1971. Synchronization of mammalian cells with respect to the mitotic cycle. *Cell Tis. Kinet.* 4:375–398.

Nishimoto, T., T. Takahashi, and C. Basilico. 1980. A temperature-sensitive mutation affecting S phase progression can lead to accumulation of cells with G_2 DNA content. *Som. Cell Genet.* 6:465–476.

Niskanen, E. O., A. Kallis, P. P. McCann, G. Pou, S. Lyda, and A. Thornhill. 1983. Divergent effects of ornithine decarboxylase inhibition on murine erthropoietic precursor cell proliferation and differentiation. *Cancer Res.* 43:1536–1540.

Nister, M., C. H. Heldin, A. Wasteson, and B. Westermark. 1984. A glioma-derived analog to platelet-derived growth factor: demonstration of receptor competing activity and immunological cross reactivity. *Proc. Natl. Acad. Sci.* 81:926–930.

Noguchi, H., G. P. Reddy, and A. B. Pardee. 1983. Rapid incorporation of label from ribonucleoside diphosphates into DNA by a cell free high molecular weight fraction from animal cell nuclei. *Cell* 32:443–451.

Norrby, K. 1970. Population kinetics of normal, transforming and neoplastic cell lines. *Acta. Pathol. Microbiol. Scand.* 70(suppl. 214): 1–50.

Norwood, T. H., W. R. Pendergrass, C. A. Sprague, and G. M. Martin. 1974. Dominance of the senescent phenotype in heterokaryons between replicative and post-replicative human fibroblast-like cells. *Proc. Natl. Acad. Sci.* 71:2231–2235.

Novi, A. M., and R. Baserga. 1971. Association of hypertrophy and DNA synthesis in mouse salivary glands after chronic administration of isoproterenol. *Am. J. Path.* 62:295–308.

——— 1972. Correlation between synthesis of ribosomal RNA and stimulaton of DNA synthesis in mouse salivary glands. *Lab. Invest.* 26:540–547.

O'Brien, T. G., M. A. Lewis, and L. Diamond. 1979. Ornithine decarboxylase activity and DNA synthesis after treatment of cells in culture with 12-O-Tetradecanoylphorbol-13-acetate. *Cancer Res.* 39:4477–4480.

Ohno, K., and G. Kimura. 1984. Genetic analysis of control of proliferation in fibroblastic cells in culture II. *Som. Cell Mol. Genet.* 10:29–36.

Okada, Y. 1962. Analysis of giant polynuclear cell formation caused by HVJ virus from Ehrlich's ascites tumor cells. *Exp. Cell Res.* 26:98–107.

Oostra, B. A., R. Harvey, B. K. Ely, A. F. Markham, and A. E. Smith. 1983. Transforming activity of polyoma virus middle-T antigen probed by site-directed mutagenesis. *Nature* 304:456–459.

Osborn, M., and K. Weber. 1977. The display of microtubules in transformed cells. *Cell* 12:561–571.

Otto, A. M., M. O. Ulrich, A. Zumbe, and L. J. deAsua. 1981. Microtubule-disrupting agents affect two different events regulating the initiation of DNA synthesis in Swiss 3T3 cells. *Proc. Natl. Acad. Sci.* 78:3063–3067.

Owens, G. K., P. S. Rabinovitch, and S. M. Schwartz. 1981. Smooth muscle cell hypertrophy versus hyperplasia in hypertension. *Proc. Natl. Acad. Sci.* 78:7759–7763.

Ozanne, B., T. Wheeler, and P. L. Kaplan. 1982. Cells transformed by RNA and DNA tumor viruses produce transforming factors. *Fed. Proc.* 41:3004–3007.

Palmiter, R. D., R. L. Brinster, R. E. Hammer, M. E. Trumbauer, M. G. Rosenfeld, N. C. Birnberg, and R. M. Evans. 1982. Dramatic growth of mice that develop from eggs microinjected with metallothionein-growth hormone fusion genes. *Nature* 300:611–615.

Palmiter, R. D., G. Norstedt, R. E. Gelinas, R. E. Hammer, and R. L. Brinster. 1983. Metallothionein-human GH fusion genes stimulate growth of mice. *Science* 222:809–814.

Pardee, A. B. 1974. A restriction point for control of normal animal cell proliferation. *Proc. Natl. Acad. Sci.* 71:1286–1290.

Pardee, A. B., R. Dubrow, J. L. Hamlin, and R. F. Kletzien. 1978. Animal cell cycle. *Ann. Rev. Biochem.* 47:715–750.

Parmley, R. T., L. W. Dow, and A. M. Mauer. 1977. Ultrastructural cell cycle-specific nuclear and nucleolar changes of human leukemic lymphoblasts. *Cancer Res.* 37:4313–4335.

Paul, D., K. D. Brown, H. T. Rupniak, and H. J. Ristow. 1978. Cell cycle regulation of growth factors and nutrients in normal and transformed cells. *In vitro* 14:76–84.

Pegoraro, L., and R. Baserga. 1970. Time of appearance of deoxythymidylate kinase and deoxythymidylate synthetase and of their templates in isoproterenol-stimulated deoxyribonucleic acid synthesis. *Lab. Invest.* 22:266–271.

Pickett-Heaps, J. D., D. H. Tippit, and K. R. Porter. 1982. Rethinking mitosis. *Cell* 29:729–744.

Pilgrim, C., W. Erb and W. Maurer. 1963. Diurnal fluctuations in the number of DNA synthesizing nuclei in various mouse tissues. *Nature* 199:863.

Piña, M., and M. Green. 1969. Biochemical studies on adenovirus multiplication. *Virology* 38:573–586.

Pipas, J. M., K. W. C. Peden, and D. Nathans. 1983. Mutational analysis of simian virus 40 T antigen: isolation and characterization of mutants with deletions in the T antigen gene. *Mol. Cell. Biol.* 3:203–213.

Pledger, W. J., P. H. Howe, and E. B. Leaf. 1982. The regulation of cell proliferation by serum growth factors. *Ann. N. Y. Acad. Sci.* 397:1–10.

Pledger, W. J., C. A. Hart, K. L. Locatell, and C. D. Scher. 1981. Platelet-derived growth factor modulated proteins: constitutive synthesis by a transformed cell line. *Proc. Natl. Acad. Sci.* 78:4358–4362.

Plumb, M., J. Stein, and G. Stein. 1983. Coordinate regulation of multiple histone mRNA's during the cell cycle in HeLa cells. *Nucleic Acids Res.* 11:2391–2410.

Pochron, S. F., and R. Baserga. 1979. Histone H1 phosphorylation in cell cycle-specific temperature-sensitive mutants of mammalian cells. *J. Biol. Chem.* 254:6352–6356.

Pochron, S., M. Rossini, Z. Darzynkiewicz, F. Traganos, and R. Baserga. 1980. Failure of accumulation of cellular RNA in hamster cells stimulated to synthesize DNA by infection with adenovirus 2. *J. Biol. Chem.* 255:4411–4413.

Pontèn, J. 1971. *Spontaneous and Virus Induced Transformation in Cell Culture* New York: Springer-Verlag.

Popolo, L., and L. Alberghina. 1984. Identification of a labile protein involved in the G_1-to-S transition in Saccharomyces cerevisiae. *Proc. Natl. Acad. Sci.* 81:120–124.

Potten, C. S., R. Schofield, and L. G. Lajtha. 1979. A comparison of cell replacement in bone marrow, testis and three regions of surface epithelium. *Biochim. Biophys. Acta.* 560:281–299.

Potten, C. S., and L. G. Lajtha. 1982. Stem cells versus stem lines. *Ann. N. Y. Acad. Sci.* 397:49–61.

Pringle, J. R. 1978. The use of conditional lethal cell cycle mutants for temporal and functional sequence mapping of cell cycle events. *J. Cell. Physiol.* 95:393–406.

Quastler, H. 1962. The analysis of cell population kinetics. In *Cell Proliferation,* eds. L. F. Lamerton and R. J. M. Fry. Oxford: Blackwell, pp. 18–34.

Quastler, H., and F. G. Sherman. 1959. Cell population kinetics in the intestinal epithelium of the mouse. *Exp. Cell Res.* 17:420–438.

Rao, P. N., and R. T. Johnson. 1970. Mammalian cell fusion: studies on the regulation of DNA synthesis and mitosis. *Nature* 225:159–164.

——— 1974. Regulation of cell cycle in hybrid cells. In *Control of Proliferation in Animal Cells,* eds. B. Clarkson and R. Baserga. Cold Spring Harbor Laboratory, pp. 785–800.

Rao, P. N., and M. L. Smith. 1981. Differential response of cycling and noncycling cells to inducers of DNA synthesis and mitosis. *J. Cell Biol.* 88:649–653.

Rao, P. N., B. Wilson, and T. T. Puck. 1977. Premature chromosome condensation and cell cycle analysis. *J. Cell. Physiol.* 91:131–142.

Reddy, J., M. Chiga, and D. Svoboda. 1969. Initiation of the division cycle of rat hepatocytes following a single injection of thioacetamide. *Lab. Invest.* 20:405–411.

Reichard, P., and R. Eliasson. 1979. Synthesis and function of polyoma initiator RNA. *Cold Spring Harbor Symp. Quant. Biol.* 43:271–277.

Renkawitz, R., H. Beug, T. Graf, P. Matthias, M. Grez, and G. Schutz. 1982. Expression of a chicken lysozyme recombinant gene is regulated by progesterone and dexamethasone after microinjection into oviduct cells. *Cell* 31:167–176.

Rheinwald, J. G., and H. Green. 1977. Epidermal growth factor and the multiplication of cultured human epidermal keratinocytes. *Nature* 265:421–424.

Riddle, V. G. H., and A. B. Pardee. 1980. Quiescent cells but not cycling cells exhibit enhanced actin synthesis before they synthesize DNA. *J. Cell. Physiol.* 103:11–15.

Ringertz, N. R., and R. E. Savage. 1976. *Cell Hybrids.* New York: Academic Press.

Robbins, E., G. Jentzsch, and A. Micali. 1968. The centriole cycle in synchronized HeLa cells. *J. Cell Biol.* 36:329–339.

Robinson, J., and D. Smith. 1981. Infection of human B lymphocytes with high multiplicities of Epstein-Barr virus: kinetics of EBNA expression, cellular DNA synthesis and mitosis. *Virology* 109:336–343.

Ronning, O. W., T. Lindmo, E. O. Pettersen, and P. O. Seglen. 1981. The role of protein accumulation in the cell cycle control of human NHIK 3025 cells. *J. Cell. Physiol.* 109:411–418.

Roscoe, D. H., H. Robinson, and A. W. Carbonell. 1973. DNA synthesis and mitosis in a temperature sensitive Chinese hamster cell line. *J. Cell. Physiol.* 82:333–338.

Ross, R. 1981. The platelet-derived growth factor. In *Tissue Growth Factors,* ed. R. Baserga. Heidelberg: Springer-Verlag, pp. 133–159.

Rossini, M., and R. Baserga. 1978. RNA synthesis in a cell cycle-specific temperature sensitive mutant from a hamster cell line. *Biochemistry* 17:858–863.

Rossini, M., G. J. Jonak, and R. Baserga. 1981. Identification of adenovirus 2 early genes required for induction of cellular DNA synthesis in resting hamster cells. *J. Virol.* 38:982–986.

Rossini, M., R. Weinmann, and R. Baserga. 1979. DNA synthesis in temperature-sensitive mutants of the cell cycle infected by polyoma virus and adenovirus. *Proc. Natl. Acad. Sci.* 76:4441–4445.

Rossini, M., S. Baserga, C. H. Huang, C. J. Ingles, and R. Baserga. 1980. Changes in RNA polymerase II in a cell cycle specific temperature-sensitive mutant of hamster cells. *J. Cell. Physiol.* 103:97–103.

Rossow, P. W., V. G. H. Riddle, and A. B. Pardee. 1979. Synthesis of labile, serum-dependent protein in early G_1 controls animal cell growth. *Proc. Natl. Acad. Sci.* 76:4446–4450.

Roufa, D. J. 1978. Replication of a mammalian genome: the role of de novo protein biosynthesis during S phase. *Cell* 13:129–138.

Rovera, G., R. Baserga, and V. Defendi. 1972. Early increase in nuclear acidic protein synthesis after SV40 infection. *Nature* 237:240–241.

Rovera, G., D. Santoli, and C. Damsky. 1979. Human promyelocytic leukemia cells in culture differentiate into macrophage-like cells when treated with a phorbol diester. *Proc. Natl. Acad. Sci.* 73:2779–2783.

Rubin, H. 1975. Central role for magnesium in coordinate control of metabolism and growth in animal cells. *Proc. Natl. Acad. Sci.* 72:3551–3555.

Ruley, H. E. 1983. Adenovirus early region 1A enables viral and cellular transforming genes to transform primary cells in culture. *Nature* 304:602–606.

Ruscetti, F. W., and R. C. Gallo. 1981. Human T-lymphocyte growth factor: regulation of growth and function of T lymphocytes. *Blood* 57:379–394.

Ruscetti, S. K., and E. M. Scolnick. 1983. Expression of a transformation related protein (p53) in the malignant stage of Friend virus-induced diseases. *J. Virol.* 46:1022–1026.

Russell, D. H. 1983. Microinjection of purified ornithine decarboxylase into Xenopus oocytes selectively stimulates ribosomal RNA synthesis. *Proc. Natl. Acad. Sci.* 80:1318–1321.

Ryan, J., P. E. Barker, K. Shimizu, M. Wigler, and F. H. Ruddle. 1983. Chromosomal assignment of a family of human oncogenes. *Proc. Natl. Acad. Sci.* 80:4460–4463.

Salmon, W. D., Jr., and W. H. Daughday. 1957. A hormonally controlled serum factor which stimulates sulfate incorporation by cartilage in vitro. *J. Lab. Clin. Med.* 49:825–836.

Sandritter, W., and G. Scomazzoni. 1964. Deoxyribonucleic acid content (Feulgen photometry) and dry weight (interference microscopy) of normal and hypertrophic heart muscle fibers. *Nature* 202:100–101.

Santos, E., D. Martin-Zanca, E. P. Reddy, M. A. Pierotti, G. Della Porta, and M. Barbacid. 1984. Malignant activation of a K-ras oncogene in lung carcinoma but not in normal tissue of the same patient. *Science* 223:661–664.

Sara, V. R., K. Hall, C. H. Rodeck, and L. Wetterberg. 1981. Human embryonic somatomedin. *Proc. Natl. Acad. Sci.* 78:3175–3179.

Sarkar, S., M. C. Glassy, S. Ferrone, and O. W. Jones. 1980. Cell cycle and the differential expression of HLA-A,B and HLA-DR antigens on human B lymphoid cells. *Proc. Natl. Acad. Sci.* 77:7297–7301.

Sasaki, T., G. Litwack, and R. Baserga. 1969. Protein synthesis in the early prereplicative phase of isoproterenol-stimulated synthesis of deoxyribonucleic acid. *J. Biol. Chem.* 244:4831–4837.

Sato, G., and L. Reid. 1978. Replacement of serum in cell culture by hormones. *Int. Rev. Biochem.* 20:219–251.

Sato, G., and R. Ross. 1979. *Hormones and Cell Culture.* Cold Spring Harbor Laboratory.

Sato, G. H., A. B. Pardee, and D. A. Sirbasku. 1982. *Growth of Cells in Hormonally Defined Media.* Cold Spring Harbor Laboratory.

Sauer, G., and V. Defendi. 1966. Stimulation of DNA synthesis and complement-fixing antigen production by SV40 in human diploid cell cultures: evidence for "abortive" infection. *Proc. Natl. Acad. Sci.* 56:452–457.

Savage, C. R., Jr., J. H. Hash, and S. Cohen. 1973. Epidermal growth factor: location of disulfide bonds. *J. Biol. Chem.* 248:7669–7672.

Savion, N., I. Vlodavsky, and D. Gospodarowicz. 1980. Role of the degradation process in the mitogenic effect of epidermal growth factor. *Proc. Natl. Acad. Sci.* 77:1466–1470.

Schaer, J. C., and U. Maurer. 1982. Thymidine nucleotide synthesis and catabolism by CHO cells and their changes during the cell cycle. *Biochim. Biophys. Acta.* 697:221–228.

Scheffler, I. E., and G. Buttin. 1973. Conditionally lethal mutations in Chinese hamster cells. I, Isolation of a temperature-sensitive line and its investigation by cell cycle studies. *J. Cell. Physiol.* 81:199–216.

Scher, C. D., R. C. Shepard, H. N. Antoniades, and C. D. Stiles. 1979. Platelet-derived growth factor and the regulation of the mammalian fibroblast cell cycle. *Biochim. Biophys. Acta.* 560:217–241.

Scher, C. D., R. L. Dick, A. P. Whipple, and K. L. Locatell. 1983. Identification of a BALB/c-3T3 cell protein modulated by platelet-derived growth factor. *Mol. Cell. Biol.* 3:70–81.

Schlegel, R., and T. L. Benjamin. 1978. Cellular alterations dependent upon the polyoma virus Hr-t function: separation of mitogenic from transforming capacities. *Cell* 14:587–599.

Schmiady, H., M. Munke, and K. Sperling. 1979. Ag-staining of nucleolus organizer regions on human prematurely condensed chromosomes from cells with different ribosomal RNA gene activity. *Exp. Cell Res.* 121:425–428.

Schneiderman, M. H., W. C. Dewey, and D. P. Highfield. 1971. Inhibition of DNA synthesis in synchronized Chinese hamster cells treated in G_1 with cycloheximide. *Exp. Cell Res.* 67:147–155.

Schreiber, A. B., I. Lax, Y. Yarden, Z. Eshhar, and J. Schlessinger. 1981. Monoclonal antibodies against receptor for epidermal growth factor induce early and delayed effects of epidermal growth factors. *Proc. Natl. Acad. Sci.* 78:7535–7539.

Schutzbank, T., R. Robinson, M. Oren, and A. J. Levine. 1982. SV40 large tumor antigen can regulate some cellular transcripts in a positive fashion. *Cell* 30:481–490.

Schwab, M., K. Alitalo, H. E. Varmus, J. M. Bishop, and D. George. 1983. A cellular oncogene (c-ki-ras) is amplified, overexpressed and located within karyotypic abnormalities in mouse adrenocortical tumor cells. *Nature* 303:497–501.

Scott, M. R. D., K. H. Westphal, and P. W. J. Rigby. 1983. Activation of mouse genes in transformed cells. *Cell* 34:557–567.

Scott, W. A., W. W. Brockman, and D. Nathans. 1976. Biological activities of deletion mutants of simian virus 40. *Virology* 75:319–334.

Seidenfeld, J., J. W. Gray, and L. J. Marton. 1981. Depletion of 9L rat brain tumor cell polyamine content by treatment with D, L-α-Difluoromethyl-ornithine inhibits proliferation and the G_1 to S transition. *Exp. Cell Res.* 131:209–216.

Sereni, A., and R. Baserga. 1981. Routine growth of cell lines in medium supplemented with milk instead of serum. *Cell Biol. Intern. Rep.* 5:339–345.

Sheinin, R. 1976. Preliminary characterization of the temperature-sensitive defect in DNA replication in a mutant mouse L cell. *Cell* 7:49–57.

Sheppard, J. R., and S. Bannai. 1974. Cyclic AMP and cell proliferation. In *Control of Proliferation in Animal Cells,* eds. B. Clarkson and R. Baserga. Cold Spring Harbor Laboratory, pp. 571–579.

Shih, C., B. Shilo, M. P. Goldfarb, A. Dannenberg, and R. A. Weinberg. 1979. Passage of phenotypes of chemically transformed cells via transfection of DNA and chromatin. *Proc. Natl. Acad. Sci.* 76:5714–5718.

Shiomi, T., and K. Sato. 1976. A temperature sensitive mutant defective in mitosis and cytokinesis. *Exp. Cell Res.* 100:297–302.

Shubik, P., R. Baserga, and A. C. Ritchie. 1953. The life and progression of induced skin tumors in mice. *Brit. J. Cancer* 7:342–351.

Sinclair, W. K. 1967. Hydroxyurea: effects on Chinese hamster cells grown in culture. *Cancer Res.* 27:297–308.

Singer, R. A., and G. C. Johnston. 1982. Transcription of rRNA genes and cell cycle regulation in the yeast Saccharomyces cerevisiae. In *Genetic Expression in the Cell Cycle* eds. G. M. Padilla and K. S. McCarty, Sr. New York: Academic Press, pp. 181–198.

Sisken, J. E., and E. Wilkes. 1967. The time of synthesis and the conservation of mitosis-related proteins in cultured human amnion cells. *J. Cell Biol.* 34:97–110.

Sisken, J. E., S. V. Bonner, and S. D. Grasch. 1982. The prolongation of mitotic stages in SV40-transformed vs. nontransformed human fibroblast cells. *J. Cell. Physiol.* 113:219–223.

Sisken, J. E., L. Morasca, and S. Kibby. 1965. Effects of temperature on the kinetics of the mitotic cycle of mammalian cells in culture *Exp. Cell Res.* 39:103–116.

Skog, S., E. Eliasson, and Eva Eliasson. 1979. Correlation between size and position within the division cycle in suspension cultures of Chang liver cells. *Cell Tis. Kinet.* 12:501–511.

Slater, M. L., and H. L. Ozer. 1976. Temperature-sensitive mutant of Balb/3T3 cells: description of a mutant affected in cellular and polyoma virus DNA synthesis. *Cell* 7:289–295.

Smith, C. J., C. S. Rubin, and O. M. Rosen. 1980. Insulin-treated 3T3-LI adipocytes and cell-free extracts derived from them incorporate ^{32}P into ribosomal protein S6. *Proc. Natl. Acad. Sci.* 77:2641–2645.

Smith, H. S., C. D. Scher, and G. J. Todaro. 1971. Induction of cell division in medium lacking serum growth factor by SV40. *Virology* 44:359–370.

Smith, J. B., and E. Rozengurt. 1978. Serum stimulates the Na^+, K^+ pump in quiescent fibroblasts by increasing Na^+ entry. *Proc. Natl. Acad. Sci.* 75:5560–5564.

Smith, J. C., and C. D. Stiles. 1981. Cytoplasmic transfer of the mitogenic response to platelet-derived growth factor. *Proc. Natl. Acad. Sci.* 78:4363–4367.

Soprano, K. J., G. V. Dev, C. M. Croce, and R. Baserga. 1979. Reactivation of silent rRNA genes by simian virus 40 in human-mouse hybrid cells. *Proc. Natl. Acad. Sci.* 76:3885–3889.

Soprano, K. J., N. Galanti, G. J. Jonak, S. McKercher, J. M. Pipas, K. W. C. Peden, and R. Baserga. 1983. Mutational analysis of SV40 T antigen: stimulation of cellular DNA synthesis and activation of ribosomal RNA genes by mutants with deletions in the T antigen gene. *Mol. Cell. Biol.* 3:214–219.

Soule, H. R., R. E. Lanford, and J. S. Butel. 1980. Antigenic and immunogenic characteristics of nuclear and membrane-associated simian virus 40 tumor antigen. *J. Virol* 33:887–901.

Sporn, M. B., A. B. Roberts, J. H. Shull, J. M. Smith, J. M. Ward, and J. Sodek. 1983. Polypeptide transforming growth factors isolated from bovine sources and used for wound healing in vivo. *Science* 219:1329–1331.

Spurr, N. K., E. Solomon, M. Jansson, D. Sheer, P. N. Goodfellow, W. F. Bodmer, and B. Vennstrom. 1984. Chromosomal localization of the human homologues to the oncogenes erb A and B. *The EMBO J.* 3:159–163.

Stancel, G. M., D. M. Prescott, and R. M. Liskay. 1981. Most of the G_1 period in hamster cells is eliminated by lengthening the S period. *Proc. Natl. Acad. Sci.* 78:6295–6298.

Stanners, C. P., and J. E. Till. 1960. DNA synthesis in individual L-strain mouse cells. *Biochim. Biophys. Acta.* 37:406–419.

Steel, G. G. 1977. *Growth Kinetics of Tumors.* Oxford: Clarendon Press.

Steglich, C., and I. E. Scheffler. 1982. An ornithine decarboxylase-deficient mutant of Chinese hamster ovary cells. *J. Biol. Chem.* 257:4603–4609.

Stein, G. H., R. M. Yanishewsky, L. Gordon, and M. Beeson. 1982. Carcinogen-transformed human cells are inhibited from entry into S phase by fusion to senescent cells but cells transformed by DNA tumor viruses overcome the inhibition. *Proc. Natl. Acad. Sci.* 79:5287–5291.

Stein, G. S., J. L. Stein, L. Baumbach, A. Leza, A. Lichtler, F. Marashi, M. Plumb, R. Rickles, F. Sierra, and T. Van Dyke. 1982. Organization and cell cycle regulation of human histone genes. *Ann. N. Y. Acad. Sci.* 397:148–167.

Steinberg, R. A., T. V. D. Wetters, and P. Coffino. 1978. Kinase negative mutants of S49 mouse lymphoma cells carry a trans-dominant mutation affecting expression of cAMP dependent protein kinase. *Cell* 15:1351–1361.

Stellwagen, R. H., and R. D. Cole. 1969. Histone biosynthesis in the mammary gland during development and lactation. *J. Biol. Chem.* 244:4878–4887.

Stetler, D. A., and K. M. Rose. 1982. Phosphorylation of deoxyribonucleic acid dependent RNA polymerase II by nuclear protein kinase NII: mechanism of enhanced ribonucleic acid synthesis. *Biochemistry* 21:3721–3728.

Stiles, C. D. 1983. The molecular biology of platelet-derived growth factor. *Cell* 33:653–655.

Stimac, E., D. Housman, and J. A. Huberman. 1977. Effect of inhibition of protein synthesis on DNA replication in cultured mammalian cells. *J. Mol. Biol.* 115:485–511.

St. Jeor, S. C., T. B. Albrecht, F. D. Funk, and F. Rapp. 1974. Stimulation of cellular DNA synthesis by human cytomegalovirus. *J. Virol.* 13:353–362.

Stoscheck, C. M., J. R. Florini, and R. A. Richman. 1980. The relationship of ornithine decarboxylase activity to proliferation and differentiation of L6 muscle cells. *J. Cell Physiol.* 102:11–18.

Strickland, S. 1981. Mouse teratocarcinoma cells: prospects for the study of embryogenesis and neoplasia. *Cell* 24:277–278.

Strong, L. C. 1977. Theories of pathogenesis: mutation and cancer. In *Genetics of Human Cancer,* eds. J. J. Muhihill, R. W. Miller, and J. F. Fraumeni, Jr. New York: Raven Press, pp. 401–414.

Stubblefield, E., and S. Murphree. 1968. Synchronized mammalian cell cultures. II, Thymidine kinase activity in Colcemid synchronized fibroblasts. *Exp. Cell Res.* 48:652–656.

Sukumar, S., V. Notario, D. Martin-Zanca, and M. Barbacid. 1983. Induction of mammary carcinomas in rats by nitrosomethylurea involves malignant activation of H-ras-1 locus by single point mutations. *Nature* 306:658–661.

Talavera, A., and C. Basilico. 1977. Temperature-sensitive mutants of BHK cells affected in cell cycle progression. *J. Cell. Physiol.* 92:425–436.

Tanuma, S., and Y. Kanai. 1982. Poly(ADP-ribosyl)ation of chromosomal proteins in the HeLa S3 cell cycle. *J. Biol. Chem.* 257:6565–6570.

Taylor, J. H. 1960. Asynchronous duplication of chromosomes in cultured cells of Chinese hamster. *J. Biophys. Biochem. Cytol.* 7:455–463.

Temin, H., and S. Mizutani. 1970. RNA-dependent DNA polymerase in virions of Rous sarcoma virus. *Nature* 226:1211–1213.

Terasima, T., and L. J. Tolmach. 1963. Growth and nucleic acid synthesis in synchronously dividing populations of HeLa cells. *Exp. Cell Res.* 30:344–362.

Terasima, T., and M. Yasukawa. 1966. Synthesis of G_1 protein preceding DNA synthesis in cultured mammalian cells. *Exp. Cell Res.* 44:669–672.

Thomas, G., G. Thomas, and H. Luther. 1981. Transcriptional and translational control of cytoplasmic proteins after serum stimulation of quiescent Swiss 3T3 cells. *Proc. Natl. Acad. Sci.* 78:5712–5716.

Thomas, G., M. Siegmann, A. M. Kubler, J. Gordon, and L. J. de Asua. 1980. Regulation of 40S ribosomal protein S6 phosphorylation in Swiss mouse 3T3 cells. *Cell* 19:1015–1023.

Thomas, G., J. Martin-Perez, M. Siegmann, and A. Otto. 1982. The effect of serum, EGF, PGF_2 and insulin in S6 phosphorylation and the initiation of protein and DNA synthesis. *Cell* 30:235–242.

Thomas, R., L. Kaplan, N. Reich, D. P. Lane, and A. J. Levine. 1983. Characterization of human p53 antigens employing primate specific monoclonal antibodies. *Virology* 131:502–517.

Thompson, L. H., C. P. Stanners, and L. Siminovitch. 1975. Selection by [^{3}H] aminoacids of CHO-cell mutants with altered leucyl- and asparagyl-transfer RNA synthetases. *Somatic Cell Genet.* 1:187–208.

Till, J. E., and E. A. McCulloch. 1961. A direct measurement of the radiation sensitivity of normal mouse bone marrow. *Radiat. Res.* 14:213–222.

Tjian, R., G. Fey, and A. Graessmann. 1978. Biological activity of purified simian virus 40 T antigen proteins. *Proc. Natl. Acad. Sci.* 75:1279–1283.

Tobey, R. A., and H. A. Crissman. 1975. Unique techniques for cell cycle analysis utilizing mithramycin and flow microfluorometry. *Exp. Cell Res.* 93:235–239.

Tobey, R. A., D. F. Petersen, and E. C. Anderson. 1971. Biochemistry of G_2 and mitosis. In *The Cell Cycle and Cancer*, ed. R. Baserga. New York: Marcel Dekker, pp. 309–353.

Todaro, G. J., G. K. Lazar, and H. Green 1965. The initiation of cell division in a contact-inhibited mammalian cell line. *J. Cell. Comp. Physiol.* 66:325–334.

Tomasovic, S. P., and W. C. Dewey. 1978. Acceleration of CHO cells into mitosis and reduction of X-ray-induced G_2 delay by cordycepin. *Exp. Cell Res.* 114:277–284.

Toniolo, D., H. K. Meiss, and C. Basilico. 1973. A temperature-sensitive mutation affecting 28S ribosomal RNA production in mammalian cells. *Proc. Natl. Acad. Sci.* 70:1273–1277.

Tooze, J. 1980. *DNA Tumor Viruses.* Cold Spring Harbor Laboratory.

Trowbridge, I. S., and F. Lopez. 1982. Monoclonal antibody to transferrin receptor blocks transferrin binding and inhibits human tumor cell growth in vitro. *Proc. Natl. Acad. Sci.* 79:1175–1179.

Tsai, Y., F. Hanaoka, M. M. Nakano, and M. Yamada. 1979. A mammalian DNA-mutant decreasing nuclear DNA polymerase activity at non-permissive temperature. *Biochem. Biophys. Res. Comm.* 91:1190–1195.

Tseng, B. Y., R. H. Grafstrom, D. Revie, W. Oertel, and M. Goulian. 1979. Studies on early intermediates in the synthesis of DNA in animal cells. *Cold Spring Harbor Symp. Quant. Biol.* 43:263–270.

Tsuboi, A., and R. Baserga. 1972. Synthesis of nuclear acidic proteins in density-inhibited fibroblasts stimulated to proliferate. *J. Cell. Physiol.* 80:107–118.

Tsutsui, Y., S. D. Chang, and R. Baserga. 1978. Failure of reactivation of chick erythrocytes after fusion with temperature-sensitive mutants of mammalian cells arrested in G_1. *Exp. Cell Res.* 113:359–367.

Tucker, R. W., A. B. Pardee, K. Fujiwara. 1979. Centriole ciliation is related to quiescence and DNA synthesis in 3T3 cells. *Cell* 17:527–535.

Türler, H. 1980. The tumor antigens and the early functions of polyoma virus. *Mol. Cell. Biochem.* 32:63–93.

Vaheri, A., E. Ruoslahti, and T. Hovi. 1974. Cell surface and growth control of chick embryo fibroblasts in culture. In *Control of Proliferation in Animal Cells*, eds. B. Clarkson and R. Baserga. Cold Spring Harbor pp. 305–312.

van Meeteren, A., G. Zoutewelle, and R. van Vijk. 1981. Differential commitment of DNA synthesis and mitosis in Reuber H35 rat hepatoma cells. *Cell Biol. Int. Rep.* 5:467.

Van Wyk, J. J., W. E. Russell, and C. H. Li. 1984. Synthetic somatomedin C: comparison with natural hormone isolated from human plasma. *Proc. Natl. Acad. Sci.* 81:740–742.

Van Wyk, J. J., L. E. Underwood, A. J. D'Ercole, D. R. Clemmons, W. J. Pledger, W. R. Wharton, and E. B. Leof. 1981. Role of somatomedin in cellular proliferation. In *The Biology of Normal Human Growth*, eds. M. Ritzen, A. Aperia, K. Hall, A. Larsson, A. Zetterberg, and R. Zetterstrom. New York: Raven Press, pp. 223–239.

Vendrely, C. 1971. Cytophotometry and histochemistry of the cell cycle. In *The Cell Cycle and Cancer*, ed. R. Baserga. New York: Marcel Dekker, pp. 227–268.

Wang, R. J. 1974. Temperature-sensitive mammalian cell line blocked in mitosis. *Nature* 248:76–78.

Waterfield, M. D., G. T. Scrace, N. Whittle, P. Stroobant, A. Johnson, A. Wasteson, B. Westermark, C. H. Heldin, J. S. Huang, and T. F. Deuel. 1983. Platelet-derived growth factor is structurally related to the putative transforming protein $p28^{sis}$ of simian sarcoma virus. *Nature* 304:35–39.

Weil, R. 1978. Viral tumor antigens: a novel type of mammalian regulator protein. *Biochim. Biophys. Acta.* 516:301–388.

Wells, D. J., L. S. Stoddard, M. J. Getz, and H. L. Moses. 1979. α-Amanitin and 5-Fluorouridine inhibition of serum-stimulated DNA synthesis in quiescent AKR-2B mouse embryo cells. *J. Cell. Physiol.* 100:199–214.

Wharton, W., E. Leaf, N. Olashaw, E. J. O'Keefe, and W. J. Pledger. 1983. Mitogenic response to epidermal growth factor (EGF) modulated by platelet-derived growth factor in cultured fibroblasts. *Exp. Cell Res.* 147:443–448.

Whelly, S., T. Ide, and R. Baserga. 1978. Stimulation of RNA synthesis in isolated nucleoli by preparations of simian virus 40 T antigen. *Virology* 88:82–91.

Widdowson, E. M. 1981. Growth of the body and its components and the influence of nutrition. In *The Biology of Normal Human Growth,* eds. M. Ritzen, K. Hall, A. Zetterberg, A. Aperia, A. Larsson, R. Zetterstrom. New York: Raven Press, pp. 253–263.

Willis, R. A., 1952. *The Spread of Tumours in the Human Body.* London: Butterworth.

Wong, T., C. Nicolau, and P. H. Hofschneider. 1980. Appearance of β-lactamase activity in animal cells upon liposome-mediated gene transfer. *Gene* 10:87–94.

Wood, J. S., and L. H. Hartwell. 1982. A dependent pathway of gene functions leading to chromosome segregation in Saccharomyces cerevisiae. *J. Cell Biol.* 94:718–726.

Wu, R. S., and W. M. Bonner. 1981. Separation of basal histone synthesis from S-phase histone synthesis in dividing cells. *Cell* 27:321–330.

Wu, R. S., S. Tsai, and W. M. Bonner. 1982. Patterns of histone variant synthesis can distinguish G_0 from G_1 cells. *Cell* 31:367–374.

Yamaizumi, M., T. Uchida, E. Mekada, and Y. Okada. 1979. Antibodies introduced into living cells by red cell ghosts are functionally stable in the cytoplasm of cells. *Cell* 18:1009–1014.

Yamashita, I., and S. Fukui. 1980. Significance of ribosomal ribonucleic acid synthesis for control of the G_1 period in the cell cycle of the heterobasidiomycetous yeast Rhodosporidium toruloides. *J. Bact.* 144:772–780.

Yarden, Y., A. B. Schreiber, and J. Schlessinger. 1982. A nonmitogenic analogue of epidermal growth factor induces early responses mediated by epidermal growth factor. *J. Cell Biol.* 92:687–693.

Yasuda, H., Y. Matsumoto, S. Mita, T. Marunouchi, and M. Yamada. 1981. A mouse temperature-sensitive mutant defective in H_1 histone phosphorylation is defective in deoxyribonucleic acid synthesis and chromosome condensation. *Biochemistry* 20:4414–4419.

Yen, A., and A. B. Pardee. 1979. Role of nuclear size in cell growth initiation. *Science* 204:1315–1317.

Yen, A., R. C. Warrington, and A. B. Pardee. 1978. Serum-stimulated 3T3 cells undertake a histidinol-sensitive process which G_1 cells do not. *Exp. Cell Res.* 114:458–462.

Yuasa, Y., S. K. Srivastava, C. Y. Dunn, J. S. Rhim, E. P. Reddy, and S. A. Aaranson. 1983. Acquisition of transforming properties by alternative point mutations within c-bas/has human proto-oncogene. *Nature* 303:775–779.

Yunis, J. J. 1983. The chromosomal basis of human neoplasia. *Science* 221:227–236.

Zaitsu, H., and G. Kimura. 1984. Arrest states in a set of mutants of rat 3Y1 cells temperature-sensitive for entering S phase. *J. Cell. Physiol.* 119:82–88.

Zavortink, M., T. Thacher, and M. Rechsteiner. 1979. Degradation of proteins microinjected into cultured mammalian cells. *J. Cell. Physiol.* 100:175–186.

Zetterberg, A. 1966. Nuclear and cytoplasmic nucleic acid content and cytoplasmic protein synthesis during interphase in mouse fibroblasts in vitro. *Exp. Cell Res.* 43:517–525.

Zetterberg, A., W. Engstrom, and O. Larsson. 1982. Growth activation of resting cells: induction of balanced and imbalanced growth. *Ann. N. Y. Acad. Sci.* 397:130–147.

Zieve, G. W., D. Turnbull, J. M. Mullins, and J. R. McIntosh. 1980. Production of large numbers of mitotic mammalian cells by use of the reversible microtubule inhibitor Nocodazole. *Exp. Cell Res.* 126:397–405.

Zouzias, D., and C. Basilico. 1979. T-antigen expression in proliferating and non-proliferating simian virus 40-transformed mouse cells. *J. Virol.* 30:711–719.

Index